The Perth N
&
Beyond

A Spirit of Enterprise & Improvement

by

Elspeth Bruce and Pat Kerr

TIPPERMUIR
· BOOKS LIMITED ·

The Perth Nursery & Beyond by Elspeth Bruce and Pat Kerr
Copyright © 2021. All rights reserved.

The right of Elspeth Bruce and Pat Kerr to be identified as the
authors of the Work has been asserted in accordance with the
Copyright, Designs & Patents Act 1988.

This first edition published and copyright 2021 by
Tippermuir Books Ltd, Perth, Scotland.
mail@tippermuirbooks.co.uk — www.tippermuirbooks.co.uk

ISBN 978-1-913836-03-0 (paperback)

A CIP catalogue record for this book is available from the British Library.

Project coordination/editiorial by Dr Paul S Philippou.

Cover design by Bernard Chandler.
Additional photographs: Roben Antoniewicz.
Editorial support: Jean Hands and Steve Zajda
Text design, layout, and artwork by Bernard Chandler [graffik].
Text set in Bauer Bodoni Std 1, 11/14pt, with Bold titling.

Printed and bound by
Park Lane Press, Leafield Industrial Estate, Corsham, Wiltshire SN13 9SW

*Dedicated to all
Beautiful Perth Volunteers
Past and Present,
and to
Barbara Bruce, 1920–2020*

'The love of gardening is natural to man.'

John Claudius Loudon
Gardener's Magazine, Volume 1 (1826)

Acknowledgements

WITH THANKS TO:

Local & Family History, AK Bell Library

Paul Adair

Roben Antoniewicz

Beautiful Perth (formerly Perth in Bloom)

Peter D A Boyd

British Newspaper Archive

Lindsay Burns, formerly of Bellwood House

Bernard Chandler

The Community Environment Challenge Fund

Culture Perth & Kinross

Christopher Dingwall

Family of WH Findlay

The Gannochy Trust

Jean Hands

Graham Hardy

The Heather Society

Family of Stan Keay

Jane Keith

The Lady Lever Art Gallery, Liverpool

Live Borders

Ishbel Mackinnon, Perth & Kinross Archive

Manchester Art Gallery

Terry Myers

The National Archives of Scotland

The National Map Library of Scotland

Perth & Kinross Council Community Greenspace Team

Perth Museum & Art Gallery

Perthshire Picture Agency

The Royal Botanic Gardens, Kew

The Royal Horticultural Society – Lindley Collection

Royal Scottish Botanical Gardens, Edinburgh

Scotland's People

Euna Scott, MBE

Nicola Small, Culture Perth & Kinross

Sir Brian Souter

Lorna Westwater, Perth & Kinross Archive

Steve Zajda

Contents

About the Authors

ELSPETH BRUCE (Secretary of Beautiful Perth) is a native of rural Perthshire and graduate of the University of Aberdeen, where she studied geography, having been inspired to do so by Kenneth McLean of Perth Academy. Graduating in 1982, she later worked in the public sector in West London, and thereafter returned to Perthshire where she now resides.

Her interest in horticulture originated from her late mother's love of gardening. Her mother was a loyal member of Perthshire Horticultural Society, winning many awards at their Biannual Show. Elspeth studied botany and was a regular visitor to the Royal Horticultural Society's garden at Wisley, Surrey, as well as the Society's Chelsea Flower Show.

She has undertaken courses in floristry and garden design. Currently, she works with over 40 communities across Perth & Kinross, supporting them with environmental improvements, such as the Riverside Park Heather Collection development.

Recognising the rich horticultural history of the area and the commercial contribution of its entrepreneurs, Elspeth felt there existed a need to further research and appreciate the subject in both a local and national context.

PAT KERR (Genealogy Researcher and Volunteer with Beautiful Perth) is a native of north Perthshire where she grew up on a Highland Pony stud with parents who were keen gardeners. On leaving Breadalbane Academy, she went up to St Andrews University but changed course to undertake the then Diploma in Primary Education.

After motherhood, another lifelong interest in history saw her becoming the honorary curator of the Clan Donnachaidh Museum, Bruar, and later a trustee and volunteer at Atholl Country Life Museum, Blair Atholl.

She assisted her husband, John, in his extensive historical research of the Atholl area, which covered many years of collecting and collating information not only from local and national records but as far as the Print Room and photographic collection at Windsor Castle. This research is now housed in the Perth & Kinross Archive and is accessible to the public.

She has undertaken museum-related courses as well as a Diploma in

Family History Studies through Stirling University. After moving to Perth, and no longer having a garden of her own, she enjoys volunteering in the Riverside Park Heather Gardens and helping to research the history of the establishment and subsequent development of the nurseries as well as the succession of people who owned them.

About Beautiful Perth

BEAUTIFUL PERTH is a local registered charity run by volunteers, in partnership with Perth & Kinross Council, its primary aim being 'making Perth a beautiful place to live, work and visit'. Beautiful Perth was first established in 1989 under the name 'Perth in Bloom' as its main focus was on horticultural works: helping produce the spectacular floral displays that have won many national and international awards over the past three decades. The name was changed to Beautiful Perth in 2007 reflecting the inclusion of environmental responsibility and community engagement in its activities.

As well as horticultural work, Beautiful Perth is involved with a broad range of community partnerships with Perth Racecourse, Police Scotland Youth Volunteers, and Turning Point, to name but a few. The charity's environmental work includes encouraging local action to improve the environment – with community litter picks and local neighbourhood improvements in locations such as Curfew Row and Craigie Cross.

Beautiful Perth runs several major projects such as the ongoing development of the Heather Collection at Riverside Park and leading the 'Zero Waste Perth' consortium. In 2018, Riverside Park was awarded the prestigious Royal Horticultural Society's 'Best UK Park and Greenspace Award'. In 2019, Beautiful Perth was named the RHS Britain in Bloom 'Champion of Champions' for the first time in its history. In the same year, Riverside Park was again voted by the RHS as the 'Best UK Park and Greenspace'.

'How appropriate that the site of the former Riverside Nursery maintains such a strong horticultural link with its significant past, and that the health benefits of greenspace, horticulture and food security remain so relevant in twenty-first century Britain.'

Beautiful Perth
October 2020

Exhibition Extraordinary in the Horticultural Room
The Royal Horticultural Society of London
(*Royal Horticultural Society Lindley Collection*)

Preface

THE TAG of the Scots people as being a 'nation of gardeners' can only be applied from around the 1750s, relatively late in comparison to the thousand or more years of garden cultivation as is the case in Syria, Iraq, Italy and Japan.

One of the earliest records of gardening in what today is Perth & Kinross is that of the royal gardens at Perth, given to the Dominican Friars on their arrival in Perth by Alexander II around 1230. These gardens faced the River Tay and the North Inch, and lay adjacent to the Castle Gavel. The friary was often occupied by kings until the capital was transferred from Perth to Edinburgh after the nearby murder of James I. It was from the monastery's spacious summer-house, called the Gilten Arbour, that Robert III witnessed the Battle of The Clans in 1396, immortalised by Walter Scott in *The Fair Maid of Perth* (1828).

In the beginning, gardening was only practised by the monastic establishments, such as at the House of the Carthusian Order in Perth, the very wealthy, and royalty from as early as the fifteenth century. Scotland had, however, established medicinal and physic gardens during the Middle Ages.

Scotland's first gardening book, *The Scots Gard'ner*, written by John Reid and published in 1683, summarised the need for order and discipline in gardens.

In eighteenth century Scotland, vast stretches of the country had become virtually treeless, as much of the harvested timber had been used for fuel and building, and little was replaced as correctly required today. At that time, much of the land was waterlogged and in need of proper drainage; little was enclosed. Laws, such as the Entail Improvement Act 1770, were passed to compel people to plant trees and hedges, but progress was slow. Ornamental gardening was more generally introduced into Scotland at the beginning of the eighteenth century with formal styles being introduced from the Continent. Land clearances started and continued on into the nineteenth century, creating large agricultural and sporting estates with few people, leaving the house or castle at their centre often still surrounded by the old medieval walled kitchen garden or physic garden.

Whilst the Napoleonic Wars (1803-15) put a halt to the original 'Grand Tour' of the previous century, once peace was restored prosperous Scots travelled far and wide inspired by plants and ideas that settled comfortably with the temperate Scottish climate. The 'English Romantic' style had been revolutionising Europe, with parkland trees, grottos, follies, and apparently natural planting using the vast lands and vistas of the great landowners.

By the nineteenth century, Scotland, along with the rest of Britain, saw increased prosperity through greater industrialisation. The Industrial Revolution saw huge profits being made by many of Scotland's entrepreneurs who, in turn, established their own townhouse and country estate gardens, allowing them to escape from the resultant squalor and dirt through the romantic and the picturesque: in art, in poetry and novels, and in the design of landscape. For many, such as hinds (farm servants), however, horticulture was a means of fulfilling the basic needs of food, with the cultivation of the cottar's garden.

This was also an age informed by the Scottish Enlightenment, an unsurpassed era of intellectual and scientific accomplishment in which a network of horticultural and scientific societies emerged – The Linnean Society of London, for example, was created in 1788. The Horticultural Society of London, now the Royal Horticultural Society, was created in 1804. Perthshire Florist and Vegetable Society, now the Royal Horticultural Society of Perthshire, was founded in 1806. The Caledonian Horticultural Society (known as the Caley) was created a little later, in 1809 and was given its first royal charter in 1820, thereafter becoming the Royal Caledonian Horticultural Society. At the outset, these were male bastions of knowledge and enlightenment, attended by landowners, head gardeners and business-men. Minutes of the first meetings reveal a tradition of producing currant wine at the Caley meetings in Edinburgh and the passing of a snuff horn at the Perth meetings held in the town's Hammerman Tavern.

The aims of the first British horticultural society, whose foundation was proposed by the horticulturist John Wedgwood, were relatively modest: to hold regular meetings which would allow them to present the society's members the opportunity to deliver papers on their horticultural discoveries and activities, to encourage discussion of the contents and publish the results. The society would also present prizes for horticultural achievement. The Caley followed similarly, with topics as diverse as the production of opium, the soporific effects of lettuce, and protecting apple blossom from the effects of spring frosts.

It was against this backdrop that the area now covered by Riverside Park, and extending outwith, that a nationally important nursery was established by James Dickson. Its notable clientele were later to include HRH The Duke of York, who sourced trees for Windsor Great Park from the nursery.

The rich alluvial deposits left by the River Tay, its sheltered position, gently undulating topography and access to transport links had created an ideal physical environment from which a plant nursery would flourish. Some areas of the site were also exposed and less sheltered, allowing tree saplings to build up strength and thus gain a reputation for reliable stock. Diversity of soil types and geology also provided an ideal basis for plant trials. The physical proximity to the estates of Scotland's wealthy landed gentry, such as Atholl, Fingask, Taymouth, Mansfield and Buccleuch was also undoubtedly another factor in its initial and ongoing success.

There was also a notable mobility of gardening labour and expertise, mostly to the south, but this strong diaspora maintained loyal connections with the north; such as John Forbes, the head gardener at Woburn Abbey, originally from St Madoes, who purchased plants from the Perth nursery.

The nursery not only sold tree stock, but also developed and sold ornamental plants and flowers. The first double-flowered Scots Roses were cultivated in the nursery by Messrs Dickson & Brown (later Dickson & Turnbull) of Perth in 1793, after Robert Brown and his brother transplanted some of the wild Scots Roses *(Rosa pimpinellifolia)* from Kinnoull Hill into the nursery. This is believed to have formed the initial basis of commercial rose breeding such as we know it today. The nursery was also the first to distribute one of the most valuable Scottish root crops – the Swedish turnip – first grown from seeds sent to the nursery by the great botanist Carl Linnaeus in 1772.

The introduction of scarlet hawthorn (*Crataegus oxyacantha 'Punicea'*) is attributed to Robert Brown, who also introduced the traditional Christmas favourite, poinsettia, to Britain from the US after retiring from the nursery and undertaking plant collecting (as had his contemporaries). This was an unprecedented era in terms of plant collecting and horticultural trends.

The nursery was also known to the notable horticulturalists and reformers of the time: John Claudius Loudon, Patrick Neill, Archibald Gorrie, Sir James Edward Smith, and John McNab.

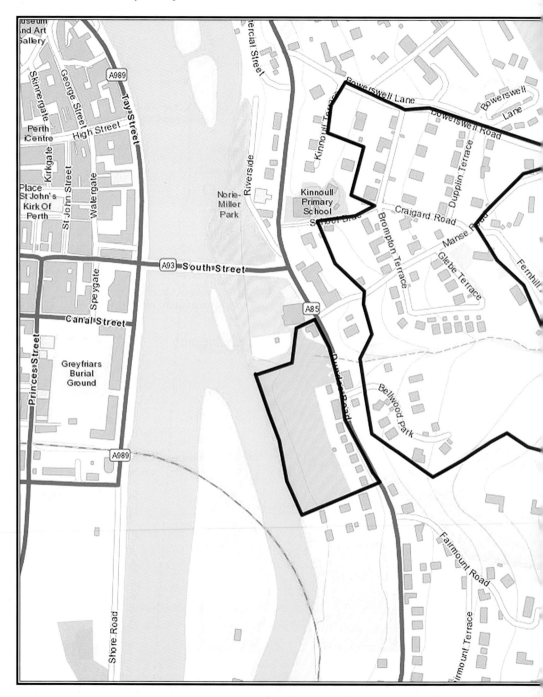

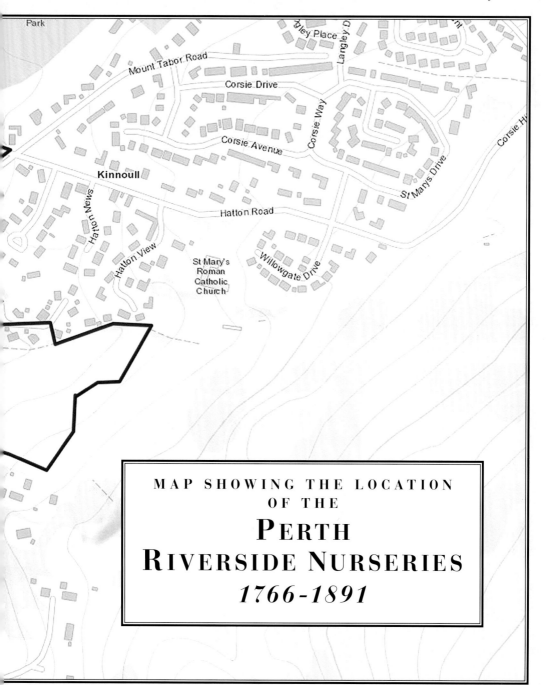

MAP SHOWING THE LOCATION
OF THE
PERTH
RIVERSIDE NURSERIES
1766-1891

Aerial View of Perth, 1940s/1950s
See Map on Previous Pages for Location of the Former Perth Riverside Nurseries

Carrots, *Daucus carota sativus*
(*Royal Horticultural Society Lindley Collection*)

Chapter 1

The Roots of the Dickson Dynasty

NURSERY ETC.— About the year 1767, a nursery was begun in this parish by Mr James Dickson from Hassendeanburn, near Hawick. He was soon after succeeded by his brother, Mr William Dickson and his present partner Mr James Brown, by whom it has been extended for upwards of 20 years, on a very extensive scale, and with that degree of success, to which their industry and taste are so justly entitled. constant employment. The soil and exposure of these grounds, are both remarkable fitted for rearing plants of such a vigorous and hardy nature, as are suitable not only to the sheltered, but to the exposed situations, which the varied face of the county abounds. This nursery contains all kinds of fruit, and forest trees, evergreen and flowering shrubs, flower roots and plants, which are naturalized to the climate. The proprietors have always been particularly careful in the selection and proof of the various kinds of fruit-trees, and in consequence of this, the demand has been very extensive. On the nursery grounds, (south from the church of Kinnoul) they have lately erected a large well-constructed green-house, in which there is not only a numerous, but a rich collection of exotic plants. And at their shop in Perth, they keep a complete assortment of all kinds of garden, tree and flower-seeds. These various articles find a ready market, not only in the rich adjacent country, but in the more remote parts of Scotland. They are even frequently sent to England and Ireland; for which, the many opportunities of water-carriage from Perth, particularly to London, and other places in south Britain, give this nursery great advantages. The happy effects of the establishment of such a branch of trade, are not confined to the actual operators, but are conspicuous on the face of the whole surrounding country. The plants and trees in our gardens and orchards, have not only increased in number, but improved in quality. Numerous and extensive plantations have been formed, and all are thriving so well, that an example productive of so much ornament and utility to the country, is every year finding many imitators.

Reverend Lewis Dunbar, *The Statistical Account of Scotland: Kinnoul County of Perth*, OSA 1796, Volume 18

'The Perth Nurseries', by Thomas Hunter
(*Courtesy of Culture Perth & Kinross*)

Robert Dickson was born in Heronhill in the Scottish Borders around 1700. He became a tenant of a small piece of land at Hassendeanburn, near Hawick, in 1728. There he established a tree nursery and became the first nurseryman who grew and retailed plants in Britain. More enlightened agricultural and horticultural ideas were beginning to emerge and in its early years, Hassendeanburn Nursery supplied mostly agricultural products and services. Such goods and services included the sale of hay, honey, beer, oats, eggs, skins, hogs and wool, and contracts for agricultural work. Between

1739 and 1766, their daybooks reveal that the entries change to sales of transplanted trees including Scots pine, fir, oak, poplar, lime, laburnum, elm, thorn, beech, ash, and fruit trees. The tree saplings were raised to help satisfy the demand from the wealthy who were undertaking grand planting schemes. Such clients included the Duke of Argyll who purchased 30,000 seedbed firs at a cost of £5. Regular orders were received from the Dukes of Roxburghe and Buccleuch, the Lords Minto and Carnworth, and the Earls of Home and Bute. Orders were also received from local people, farmers, and gardeners with regular purchases of 'cabitch plants, purtates, onions and leeks' and seeds of 'hearbs'.

The Dickson family were specialists in trees, but they were also pioneers of nursery stock and a profitable seeds industry. Robert Dickson also acted as a moneylender in Hawick before any banks were established in his local town.

From Hassendeanburn, trees could be sent out both north and south, which meant that the nursery at that time dominated the markets of Scotland and the north of England. The Dicksons were fiercely competitive businessmen, undercutting the prices charged by their English rivals, facilitated by the Dickson's large scale production on relatively economical land and employment of cheap labour.

There are a number of accounts of the Dickson family, but the most reliable sources are their wills and official records, upon which this account is based. Robert's son Archibald, born in 1718, had ten children – five sons and five daughters.

Archibald soon expanded his father's trading, fuelled by the ongoing demand from large estates. Much of this trade dealt with forestry and fruit trees, and the firm's daybook also indicates the beginnings of commercial trading in larch.

By 1766, the Dicksons could afford to expand by feuing 36 acres of land at Hawick to open another new nursery, hiring in extra men and apprentices. The company had now become known nationally for the excellent quality of its trees and its comprehensive selection of seeds. This was an emerging new market, with the supply and demand driving up margins, and creating healthy profits. The success of Hassendeanburn allowed expansion to enable another four nurseries belonging to the Dickson family, located at Brechin, Edinburgh, Chester, and Perth.

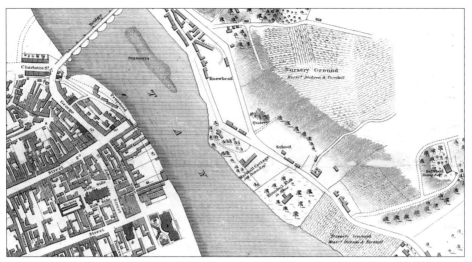

Map Showing Dickson & Turnbull Perth Riverside Nurseries, 1860
(*Reproduced with Permission of The National Library of Scotland*)

Plant nurseries had a reputation for paying their employees poorly. Dicksons, however, could afford to pay their workers between £5 and £6 per year, with allowances for shoes, oatmeal, and beer for men, and an apron, some wool and linen for women.

Two of Archibald's sons established their own nurseries, doubtlessly having identified opportunities further afield with their wealth of arboricultural and horticultural experience. Walter Dickson founded the famous firm of Dickson & Company, in Leith Walk, Edinburgh; and in 1766, the Perth Nurseries were founded by James Dickson, trading under the name of Dickson & Brown (James Brown). James Dickson was shortly succeeded by his brother William. In the historical context, this was the year in which Bonnie Prince Charlie became the new Stuart claimant to the throne, and the year in which Smeaton's Bridge was built in Perth. In *Perth, Scotland, Survey of Inhabitants 1766*, the population is recorded at 7,542. Employment was noticeably related to weaving, whilst other occupations included millers, tanners, maltsters, masons and brewers. Excise officers and surgeons are listed amongst other professions.

The following year, the building of Edinburgh's New Town commenced. The gentrification of Perth took place some 30 years later with the building of Rose Terrace and Marshall Place.

Rose Terrace, Perth (*Roben Antoniewicz*)

Marshall Place, Perth (*Roben Antoniewicz*)

Records indicate that in 1793, the brothers Robert and James Brown collected wild roses from Kinnoull Hill and propagated them on, forming the basis of commercial rose breeding, one of the horticultural introductions for which Dickson & Brown became notable.

Robert Brown retired from Dickson & Brown in 1820, whilst little is known of James. The reins were then handed over to William's nephew, Archibald Turnbull, who had arrived at Perth from Hassendeanburn in 1803. William died in 1835 leaving Archibald to succeed him as sole partner in the Perth Nursery which by then traded under the name of Dickson & Turnbull.

Archibald Turnbull of Bellwood, Perth
Hugh Thomson, c.1865, Papers of William Carmichael McIntosh
(Courtesy of the University of St Andrews Library, ms37103-4-25v)

In William Dickson's will, his property of Bellwood at Kinnoull was left to his nephew Archibald. Dickson's assets, excluding the property of Bellwood, were of considerable worth – in the region of £1.2 million at today's value. The trustees included:

Archibald Dickson of Housebyres, my Brother, Walter Dickson Nursery and Seedsman in Edinburgh, also my brother, Walter Dickson Junior, Writer to the Signet my Nephew, James Paton of Glenalmond, James Murray Patton of Glenalmond. William Gloag, Collector of Cess for the County of Perth, Archibald Turnbull, Nursery and Seedsmen in Perth, my Nephew and Robert Hope Moncrieff, Writer in Perth. . .

I do therefore, and for the love and favour and affection which I have and bear to my relations after named direct my Trustees in the first place to convey and make over to and favour of my Nephew Archibald Turnbull the lands of Barnhill, Knowhead, Feumuir, commonly called Mountabour and my whole other property in the Parish of Kinnoull subject to life rent afore mentioned. . .

In the Second place to convey and make over to and in favour of my Nephew William Dickson, son of my Brother Archibald Dickson, the lands of Feu and the properties in the High Street, Blackfriars and Speygate of Perth. . . and having been at great trouble and expense in the establishment of the Business of Nursery and Seedsmen, which I now carry on in Perth and Brechin in Copartnery with my Nephew the said Archibald Turnbull from which I have received much assistance and having embark'd a great part of my means in the advantageous prosecution of that business.

I do hereby in the third place direct my said Trustees to convey and make over to the said Archibald Turnbull and William Dickson equally between them, the whole of the share of my stock in the concern of Dickson and Turnbull belonging to me at the time of my death.

In the fourth place, I direct my Trustees to use every indulgence in the recovery of the outstanding accounts, and so as no unreasonable cause of offence may be given to customers of the business, and I would recommend to my Trustees and my Nephews or such of them as continues the business to enter into some amicable arrangement, whereby the outstanding accounts may be transferred over to my said Nephews and the levying of the amount put within their own discretion.

In the fifth place I direct my Trustees to settle and secure my sister Mrs Janet Dickson or Clark in the liferent right, use and enjoyment of the premises in Bridgend enumerated fifth…

In the sixth place I direct my Trustees to make payment to my Brother Walter and to each of my Nephews and Nieces who shall survive me (excepting Robert Dunlop, William Dunlop and the said Archibald Turnbull and William Dickson) the sum of Fifty Pounds sterling and to my said Nephews Robert and William Dunlop One Hundred Pounds Sterling each and that in token regard for them, and which mark of my regard I would have enlarged, did I not consider that the ulterior purposes of this writing will enable any residuary Legatees to provide for those of my relations who may most require it.

In the seventh place, I direct my said Trustees after fulfilling the above purposes to divide and settle the residue of my Estate in manner following, vizt.,

Dickson makes equal financial provision from the residue of his estate in consideration of his three sisters, and an annuity of £50 to his sister Mrs Clark. The contents of Bellwood were left to his nephew William Dickson, but purchasable by Turnbull on payment of £300 to William.

Of the debtors detailed in his will, Dickson wrote:

The above sum of £944.8.11d is considered wholly irrecoverable [around £100,000 today].

A great number of the parties are poor and bankrupt and those that are not so, dispute liability on the grounds of previous payment or compensation by counter claims. And the accounts all being prescribed, no value whatever can be put on the above amount.

(National Records of Scotland, SC49/3/25)

Gardens at Bellwood House, Perth (Elspeth Bruce)

A number of his properties were listed for sale within a year or so of his death including the quarry at Muirhall, residential properties in King Street and Bridgend and a half-share in commercial premises in Canal Street and Horsecross in Perth, possibly to clear the bad debts which the nursery was carrying and/or to reinvest elsewhere.

Several pieces of land in the vicinity of the Dundee Road were also offered for sale, having been left to Turnbull.

> The adjoining nursery, now the property of Archibald Turnbull Esq, has been famed throughout Britain during the last half century, for the culture of fruit and forest trees, ornamental shrubbery, and flowers in endless variety. Vast quantities of plants are annually sent from this Nursery to England. The late Duke of York, whose pleasure grounds procured the greatest number of his ornamental shrubbery from this quarter...
>
> George Penny, *Traditions of Perth* (1836)

Archibald Turnbull was also a remunerated director of five of the newly fledged railway companies and sat on many committees to support investment in the same. Annual remuneration for each directorship was fixed at £500 per annum. Turnbull invested £2,500 in each of the five rail companies. He was a director of Standard Life Assurance and was elected City Treasurer for Perth in 1829. He was a JP, a city councillor, was representative on a number of local committees, including the Inches Committee, the Cess Committee, and the Railway Committee. In 1836, he was presented with two silver claret jugs by the Perthshire Farming Association 'as a mark of respect and their high estimation of him'.

In 1840, Turnbull was elected to the Wellington Testimonial Committee – a committee formed to approve of the erection of an equestrian statue of the Duke of Wellington in Edinburgh. This statue was installed outside Register House in 1852 and is regarded as one of the best of the Duke. Wellington was depicted on his favourite horse 'Copenhagen'. In the same year as Turnbull was elected to the Wellington Testimonial Committee, a dinner was held by the Royal Perthshire Horticultural Society within the George Hotel in honour of Turnbull, as a 'mark of the high sense they entertain of his very zealous and efficient services as Honorary Secretary of this Society'. He had been a promoting member of the Royal Horticultural Society of Perthshire and regularly supported their annual shows by providing displays from his nursery. Dickson & Brown also donated a snuff horn, with horticulturally themed accessories, mounted with a silver insignia to be passed around the Royal Horticultural Society of Perthshire meetings.

> Dickson and Turnbull exhibited *Manetta cordifolia; Hydrolea spinosa;* and a beautiful assortment of Fuchsias and Liliums; and a great variety of other fine plants, along with some fine Dahlias, Verbenas &...
>
> *The Dundee Courier,* 1847

Turnbull used his connections and influence to bring about change and improvements in Perth. He was a subscriber to the Perth City and County Infirmary (now the A K Bell Library) contributing to its funds, as well as successfully persuading the Committee of the Destitute Fund that claims for relief for the city and county of Perth must be considered. He offered his premises at Horsecross as a fever ward for the City Infirmary during a typhoid epidemic. His offer was declined as the infirmary's constitution did not permit it. In 1842, he was elected treasurer to the committee which raised funds for a memorial for the ill-fated explorer David Douglas, appealing to many individuals and public bodies before successfully reaching the required target which enabled Douglas's memorial to be erected in the graveyard at New Scone where it stands today. In 1861, Turnbull chaired the inaugural meeting of The Perth Gas Consumers' Company in response to 'exorbitant' gas prices in the city.

Amidst his public duty and business interests, Turnbull enjoyed the gentlemen's leisure pursuits of horse racing, horse riding, fishing, curling, cricket, and golf. A 'Perth Nursery' handicap featured at Perth Races. He held a game certificate and was a member of the Lodge St Andrew No. 25, the Library and Antiquarian Society of Perth, and was also a committee member of the Royal Perth Golfing Society.

In his garden at Bellwood stood two cannons which had been installed by former resident Captain Young of the Kinnoull Rock Artillery. Turnbull fired the cannon on a number of occasions including royal visits by HM Queen Victoria in 1842 and 1848 and the opening of the Perth Railway Bridge in 1863.

> It would, however, be unpardonable did we omit to mention the highly seasonable aid received on the occasion from Archibald Turnbull Esq. of Bellwood. This gentleman, in the kindest and most patriotic spirit, gave the servants of the railway Directors unlimited access to his shrubberies and flower gardens, and furnished them to repletion with floral ornaments of every description. The consequence was, that these parties were enabled to make such a display of taste and elegance as has not been surpassed in our locality for a longtime past...
>
> As the hour approached for Her Majesty leaving the George Hotel, the bells of St John's and St Paul's churches commenced ringing merry peals, and the guns from the Bellwood battery fired a royal salute...

Perthshire Advertiser, 5 October 1848

Turnbull was later received at Perth Railway Station by Queen Victoria where he presented her with a bouquet of flowers.

The cannons were later donated to Perth Museum & Art Gallery. The castellated battlements are still evident within the beautiful grounds of Bellwood, where vestiges of the Perth Nursery tower towards the sky through the survival of some prominent trees.

Turnbull characterised the very epitome of enterprise and improvement and appeared to be of high moral standing. Under his direction, the nursery continued to trade very successfully, underpinned by the development of the railway transport epoch where he played his part in enjoying the financial success of his investments in the railways and the transportation efficiencies which the railways brought. Turnbull, however, was not spared the nuisance of crimes and civil debts.

In 1844, John McIntyre, a 'stone breaker' was imprisoned for one month for being equipped for poaching at Turnbull's farm of Hatton at Kinnoull. In 1862, incendiary fires at Turnbull's premises at Hatton and Bridgend were reported.

It is hoped that the perpetrator of these daring outrages will not long escape detection. Much sympathy is expressed for Mr Turnbull, who is a gentleman universally respected in city and county for his public spirit, kindness and urbanity. He employs a large number of work people, who hold him in the highest estimation.

Dundee, Perth and Cupar Advertiser, 2 January 1862

In 1870, the case of Gray versus Turnbull was heard in the House of Lords. The action was brought about by Turnbull to have it declared that a piece of land about eight yards square belonged to him and not to Gray, whose land marched on to Turnbull's.

The smallness of the matter at stake provoked the usual satirical remarks from Lords Westbury and Chelmsford about the litigiousness of the Scotch.

Dundee Courier, 23 June 1870

Turnbull died at his home of Bellwood on 19 January 1875, aged 85 years. His nephew, and manager of the Perth Nursery, John A Anderson of St Albans Cottage, Perth, was in attendance at the time of his death.

The death of Mr Archibald Turnbull deprives Perth of an old familiar figure. Mr Turnbull had for many years been as an institution. He and his dog were known to all of us. George Street was quite imperfect without 'Turnbull of Bellwood.' With the single exception, I may say of Sir David Ross, Mr Turnbull dwarfed all his neighbours and fellow citizens. When many of the men who are now old at sixty were in their boyhood, Mr Turnbull was in his the prime of early manhood. In his youth, he was a remarkably fine man, not stout, but slight and stately. He had the reputation of being one of the finest, if not the very finest horseman in Perthshire. He rode to hounds upon a little Irish horse, famous at that time; and upon this horse he performed some exceedingly neat and daring feats in equestrianship. There is, I believe, a certain stupendous, albeit a breakneck water jump near Stanley, which is known even to this day as 'Turnbull's Leap.' Altogether, Archibald Turnbull, as I remember him forty years ago, was a brave and gallant gentleman, who carried himself with much dash and aplomb. Indeed, Perth in those days was a much gayer place than it is today. The magistrates were gentlemen, and the Town Council was composed of men who imparted dignity on their office. I fear me that is a good deal changed nowadays although I am bound to say there are exceptions.

Sometimes we hear a cogent speech
Fit for a judge and jury
By far this great majority
Are useless sound and fury.

But to return to Mr Turnbull. He had the reputation of being an 'aristocrat' in manner. He certainly was 'thoray,' and this, I daresay, accounted for many other traits with which he was credited. As you know, Mr Turnbull died unmarried. Nobody could understand this but I think myself correct in saying it was due to some love affair in the bright strong days now forgotten. Mr Turnbull had long outlived the gossips of that day. Yet the story has managed to cling like lichen to the old stone walls of time, though it is now grey and old. His recollections have died with him. His end was swift at last, another tribute to the killing force of this stark and terrible winter.

'Local Chit Chat' in the *Perthshire Advertiser*, 1875

Turnbull left the Bellwood property to his niece, Jessie Dickson Anderson:

[And] from the many kindnesses and attentions shown to me during my lifetime ... make over to her the whole household furniture, silver plate and other articles thereon contained in.

Also my Horses, Carriages and Cows that may be there at my death and further desire that she may be in possession of one half of my Nursery and Seed Stock and thereafter to receive one half of the profits of the business as now carried on by me as Nursery and Seedsman so as to enable her to keep up the Establishment at Billwood and become a partner with her brother, John A Anderson. My Trustees or Trustee will make over to my Nephew John A Anderson residing at St Albans Cottage, the other half of my Nursery and Seed Stock and the other half interest in the business as carried on by me as Nursery and Seedsman and having bought St Albans Cottage which cost me upwards of Two Thousand Pounds Stg. I caused it to be made over to him (my stating this is to show what I have done for him).

I further desire that the Trustees will convey over to Jessie D Turner and John A Anderson for their joint propertys my Granaries at Canal Street and Horse Cross, the packing shed at Kinnoull Church and my house at Brechin, they being all necessary for their business.

On the death of my late Uncle I became the Proprietor of the Granarys etc at the Thornhead but to give my Nephew John A Anderson a vote in the City I caused my man of business to make over the titles in his name but have drawn the rents of it ever since. In case he should make any claim for bygone rents drawn by me, I thereby withdraw his name from any benefit from my Estate.

My two fields at Mountaber you will convey to John A Anderson, my lease of Hatton to be made over to Jessie D Turner and John A Anderson as it is of great use for the carrying on of their business, also the stocking and implements on said farm.

22 March 1875

(National Records of Scotland, SC 49/31/99)

Both William Dickson and Archibald Turnbull were laid to rest at Kinnoull Churchyard, where their graves are tended by volunteers from Beautiful Perth.

William Dickson of
Barnhill & Bellwood (*d.*1835)
Headstone in Kinnoull Churchyard
(*Elspeth Bruce*)

Archibald Turnbull of
Bellwood (*d.*1875)
Headstone in Kinnoull Churchyard
(*Elspeth Bruce*)

Chapter 2
The Tendrils of Enterprise

HASSENDEANBURN NURSERY, the original Dickson nursery, was also where John Veitch (born in 1752) was apprenticed between 1766 and 1768. After his apprenticeship, Veitch left Leith Harbour with ten shillings in his pocket and his father's blessing. He was joining the increasing diaspora of Scottish gardeners relocating south of the border. Also of Scottish descent was Philip Miller (1691-1771) the curator of the Chelsea Physic Garden, Britain's first Botanic Garden; as was William Aiton (1731-1793) the curator of Kew Botanic Gardens (between 1754 and 1793) to name only two of many. English estate gardens were becoming increasingly overseen by Scots, about which the English garden designer Stephen Switzer made the observation:

> There are several Northern Lads, which whether they have serv'd time in the Art, or not, very few of us know anything of; yet by the help of a little Learning, and a great deal of Impudence, they invade these Southern Provinces; and the natural benignity of this warmer climate has such a wonderful influence on them, that one of them knows (or pretends to know) more in one twelve-month, than a laborious, honest, South Country man does in seven years.
>
> Stephen Switzer, *Ichnographia Rustica or the Nobleman's Gentlemen's and Gardener's Recreation, Volume 1* (1717)

The simmering resentment against the Scots gardeners came to a head when Scotsmen also began to make a success of nursery gardening in and around London, beginning around 1760. The English then tried to resuscitate an old Chartered Company of Gardeners under a charter of James I, whereby a resolution was passed that no apprentice should be employed from the North.

Sue Shephard, *Seeds of Fortune: A Great Gardening Dynasty* (2003)

Veitch created the Exeter based firm of Veitch Nurseries, initiating the origins of the Veitch dynasty. He died in 1839. The Veitch Memorial Medal is now awarded annually by the Royal Horticultural Society:

> [To] persons of any nationality who have made an outstanding contribution to the advancement and improvement of the science and practice of horticulture.

Such was the Veitch dynasty's contribution to horticulture.

An unrelated James Dickson also ventured south and became a founding member of The Linnean Society of London in 1788, and in 1804 was a founding member of the Horticultural Society of London. The Linnean Society of London is the world's oldest active biological society, taking its name from the Swedish naturalist Carl Linneaus, the 'father of modern taxonomy', who formalised binomial nomenclature, used to this day.

This era was revolutionary, underpinned by land reform and networks of newly founded horticultural and agricultural societies at both local and national level. It was the age of the 'Agricultural Revolution', industrialisation, improved transport networks, plant crazes, and avid plant collecting.

Another eminent horticultural Scot was farmer's son John Claudius Loudon, born in 1783 at Cambuslang in Lanarkshire. He was apprenticed at Dickson & Shade's Edinburgh nursery, and studied at the University of Edinburgh, before departing for London in 1803 aged 20. It was in London that he met Sir Joseph Banks, the renowned botanist, naturalist, and explorer and also Jeremy Bentham, philosopher, jurist, and social reformer.

Around 1803, Loudon published his first article, proposing improvements to London squares and also made proposals for the 3rd Earl of Mansfield's park and gardens at Scone Palace, Perth. This had been preceded by the 2nd Earl of Mansfield commissioning Thomas White Senior to 'embellish the Estate'. According to Thomas Hunter, in 1883: 'his idea of embellishment was close to Vandalism, as he executed his commission by cutting down most of the old oak on the property'. Loudon's executed plans for Scone Palace included relocating the village and inhabitants of Old Scone. He became a reformer, urging on in his readership the need to rationalise their estates, their finances, and their communities. Rooted in his life-long habit of working through the night, his output was exhausting.

He also urged higher wages for gardeners and better living conditions. In 1822, Loudon published his best-known book, *The Encyclopaedia of Gardening*. In 1826, he launched the pioneering *Gardener's Magazine*, as a forum for professionals. The magazine pressed the professional gardener to raise himself by self-education.

John Claudius Loudon (1783-1843)

In 1832, Loudon established the design theory 'Gardenesque' whereby attention was given to the individual plant and placement in the best conditions for them to grow to their potential. 'Gardenesque' offered the opportunity to introduce exotics into gardens.

As arguably Britain's most prolific horticultural journalist, we are indebted to Loudon for his exhaustive list of publications in providing us with an invaluable insight into the horticultural heritage of Perth & Kinross and that of the Perth Nursery. Likewise, the writings of Patrick Neill provide a valuable insight as to the nineteenth century horticultural landscape in Scotland.

In his *On Gardens and Orchards of Scotland, Drawn up, by desire of the Board of Agriculture* (1818), Patrick Neill observed:

> Scotland has long been remarkable for producing great numbers of professional gardeners; more perhaps than any other country of Europe, of equal extent: and while, in times past, they have been numerous, several of them have also been eminent, and have attained the highest stations in their profession, not only in Great Britain, but in various foreign lands. At the present day, most of the principal nobility and gentry in England have Scottish head gardeners, while the numbers of an inferior class to be found in South Britain is quite surprising. It may not be easy to assign all the causes which have tended to produce the fame and the numbers of Scottish gardeners; but some of them may be pointed out.

Neill attributed their success to the establishment of parochial schools, and the 'diffusion' of education in general amongst the 'common people', instruction of apprentices in the evening on matters of plant nomenclature, and the close connection between gardening and the 'medical art'. The practice of herbalism also made them valuable members of their communities. Neill described their role as the 'skilly man' of the district, 'generally applied to when the operation of blood letting was to be performed' – a thriving practice in the nineteenth century. He also observed:

> [Their] daily opportunities of consulting with their employers, and the visitors of their employers, and a frequent and very commendable practice of masters indulging gardeners with access to their libraries. The gentle nature of their business, too, with the retirement which it implies, may have some effect in forming the minds of gardeners more for the study of such subjects, than those of most other of the classes in the same rank of life.
>
> They are a hardy race, accustomed to labour, and able to undergo great fatigue, often while subsisting on fare, which would not only be thought of as homely, but scanty, by their brethren in England.

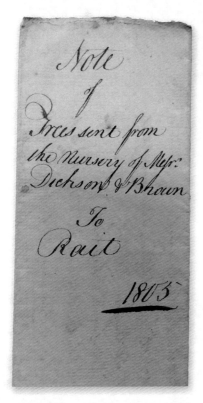

Note of Trees Sent to Rait, 1803
(*Perth & Kinross Archive*)

The Scots' ability to withstand the challenges of Scotland's 'very variable and unsettled climate' is also witnessed by Neill in his explanation of the successes of Scottish gardeners at home and abroad.

EHM Cox, however, in *A History of Gardening in Scotland* (1935) did not entirely agree with Neill's reasoning as to the emigration, and explained that an employer often preferred to 'obtain his labour from a less happy climate than his own'. It was also becoming a 'commoner occurrence for landowners to find large land-owners owning estates on both countries'. Cox further attributed the great influx of Scottish gardeners to England as originating from the time of Philip Miller, a Scot, becoming curator of the Chelsea Physic Garden; given by Sir Hans Sloane in 1722 to the Apothecaries Company. Miller preferred for apprenticeships his own countrymen who moved from the North. There were also greater opportunities in the South and it is also noted that the Scots, having left the country to work elsewhere are 'nationally minded' with a camaraderie 'which inclined them to assisting each other'.

As regards the many debatable factors behind Scottish gardeners migrating elsewhere, what is concluded, without any doubt is that between 1700 and 1850 they made a major contribution to horticulture and sustained a strong network throughout Great Britain and beyond.

The archives of the Linnean Society reveal that in 1814 Patrick Neill visited the Doo Hillock Nursery outside Forfar with Robert Brown of the Perth Nursery. Doo Hillock was at that time one of the few nurseries which supplied herbaceous plants. In a letter written by Neill to Sir James Edward Smith, (BG 110/JES/COR/7/103) he gave thanks to the Linnean Society for a donation of £31 for George Don's family.

George Don Senior (1764-1814) was apprenticed at Dupplin Castle in Perthshire. He explored the Ochil Hills and Grampians on his days off in search of plants. In 1797, Don leased two acres of land north of Forfar where he established his nursery. Don was appointed as head gardener at the Royal Botanic Gardens in Edinburgh, leaving his elderly father in charge of Doo Hillock. From Edinburgh, Don explored the surrounding district with his friend and fellow botanist Patrick Neill, as well as attending medical classes at Edinburgh University.

Don returned to Doo Hillock in 1806, resuming care of his market garden and nursery. However, in searching for wild plants, he apparently neglected his business, requiring he and his family in 1812 to apply for credit to keep the business afloat. By 1813, they were relying on neighbours for sustenance. Don died on 14 January 1814. Neill helped raise funds to help Don's widow Caroline, and allow her an opportunity to sell the nursery plants and transfer the tenancy of Doo Hillock. Of George and Caroline's 15 children, the five surviving sons were all taught gardening and followed the trade.

In a communication, Neill commented on the opposing characteristics of Don's two grown up sons: the elder, George Don (1798-1856) is 'senseless, stubborn, unfeeling', whilst the younger (David Don 1799-1841) is 'pliable, full of attachment...and really clever'. 'The Committee failed to convince the eldest to carry on the garden' and Neill feared that it would be abandoned and that the son will go into the 'army, navy or worse'. He continued to write that he will get the younger son into a nursery or into Kew.

The young George Don worked for Messrs Dickson at Broughton before moving south to the Chelsea Physic Garden. From 1821 to 1823, he collected plants in Madeira, Sierra Leone, Sao Tome, and the West Indies for the Horticultural Society of London. In addition, David went to work for Messrs Dickson at Broughton and then the Chelsea Physic Garden, before moving to become keeper of Aylmer Bourke Lambert's library and herbarium. In 1821, he accompanied his friend Patrick Neill to Paris where they met many prominent scientists of the age.

As a protégé of Sir Joseph Banks, David later became librarian to the Linnean Society, and from 1836, he was professor of botany at King's College, London. A carved plaque remains at 32 Soho Square, London, – the site of Banks's former home and meeting place of the Linnean Society between 1821 and 1857 – commemorating Sir Joseph Banks, Robert Brown, and David Don.

Sir Joseph Banks (1743-1820), by Thomas Phillips
(*Royal Horticultural Society Lindley Collection*)

Montrose born Robert Brown, FRSE, FRS, FLS, MWS (1773–1858) was another famous botanist and plant collector who is noted for describing what became termed Brownian motion. (He is not Robert Brown of Dickson & Brown.)

In conclusion, the Scottish diaspora of the era were very significant 'founding fathers' of improvements in horticulture and arboriculture, commercially and domestically. Many made personal sacrifices, and in the case of David Douglas and John Jeffrey, both of Perthshire, they paid the ultimate price in pursuing their botanical careers.

Fortunella margarita, Annona reticulata,
Prunus domestica, Citrus sinensis
(Royal Horticultural Society Lindley Collection)

Chapter 3

Horticulture in Nineteenth Century Scotland

T HE PERTH NURSERY undoubtedly peaked in its successes in the nineteenth century. The business acumen of the Dicksons and Archibald Turnbull is unquestionable, as were the botanical interests and horticultural knowledge of Robert Brown. Various archives and historical accounts emphasise the role of the landed gentry in contributing to demand for trees, ornamentals, fruit and vegetable seeds and plants, as well as sundries, however, clientele were not just restricted to the landed gentry as the list of Dickson's debtors confirms.

It is useful to examine the nature of horticulture across Scotland at this revolutionary time. *On Gardens and Orchards of Scotland: Drawn up by desire of the Board of Agriculture* (1818) was written by Patrick Neill, LLD, FRSE, FSA Scot, FLS (1776-1851), a printer, horticulturalist, and naturalist of note. Neill was also a founding member of The Caledonian Horticultural Society, which was formed in 1809 and underpinned horticultural improvements across Scotland. (Robert Brown and Archibald Turnbull both became members of the The Caledonian Horticultural Society.) Neill held the position of Secretary with the Society for a term of 40 years. He was Vice President of the Botanical Society of Edinburgh, and Secretary of the Wernerian Natural History Society (an offshoot of the Royal Society of Edinburgh). He directed the planting of West Princes Street Gardens in Edinburgh with 77,000 trees and shrubs. John Claudius Loudon described Neill's garden at Canonmills, Edinburgh, as the 'richest urban garden in the country'.

In his writings, Neill has provided a detailed and valuable overview and an insight into the socio-economics of the time. He described several categories of gardens:

1. The gardens of the noblemen or gentlemen of great landed property.

2. Market gardens in the vicinity of cities, towns, or villages. These are generally occupied by experienced gardeners, whose livelihood depends upon the produce of their gardens; and they are, therefore, well kept, and highly productive.

   [Market gardens in Perth were numerous and the produce consisted chiefly of savoys, cabbages, Scots kail, onions, early potatoes, turnips, carrots, and pease; they were located on the outskirts of Perth, albeit mostly in Barnhill. The establishment of The Depot (today HM Prison Perth) for Napoleonic prisoners of war also provided produce for Perth at this time. Built in 1812, at a cost of £130,000, it was planned to contain 7,000 men, and was 'considered to be the finest specimen of this species of architecture in Scotland'. It was occupied by the Napoleonic prisoners of war for barely two years, Bonaparte having been exiled to Elba, and the inmates released to return home.]

3. Farmers' gardens. These in general, cannot be boasted of. There are, however, some honourable exceptions.

4. The gardens of feuars, or small proprietors; some of which are very well attended to.

5. Gardens, or small yards, rented by the year by day labourers, carpenters, smiths, tailors and others, who from the nature of their engagements, have the command of a portion of their time, to perform little pieces of necessary work in their plots.

6. Kail-yards, occupied by hinds, or farm-servants. These are the lowest in the scale, and have probably undergone no improvement for centuries.

7. Village Gardens; in some cases much neglected, and in others, well kept and productive.

HM Prison, Perth (*Roben Antoniewicz*)

Neill was complimentary in his comments regarding The Royal Botanic Garden in Edinburgh which must 'stand by itself' but clearly dismissive of Glasgow Botanic Garden, which he described as 'unworthy of that opulent city'.

He continued by describing public nurseries which have 'long been established in Scotland, and have met with that ample encouragement, to be expected in an improving country. Of late years, they have become very numerous, there being hardly any considerable town in Scotland where a nursery has not been formed'. He observed that private nurseries were required, and were increasing in number to satisfy the demand for timber supplies. He furthermore noted that 'as to public or sale orchards, there are but a few plantations of fruit trees in Scotland', with the demand for 'cyder and perry' being very limited. Private orchards, however, he noted were very numerous, with those who did not possess separate orchards dedicating a portion of the kitchen garden to standard apples, pears, cherries, and plums.

Neil interestingly described Scottish cottage gardens in terms of their healthy produce from their kail yards in comparison to the Highlander's diet:

> The most eminent physicians of the last age were in use to ascribe a remarkable improvement in the health of the Scottish peasantry, – the alleviation of some prevailing diseases, and the extirpation of others, – is the general introduction of the cottage gardens, and consequent increased use of coleworts [kale], cabbages and potatoes. The Highlanders still live very much on fish, flesh, and the produce of their cattle with little vegetable ailment, excepting oatmeal and potatoes. They are now, however, becoming sensible of the value of other garden-stuffs, which not only renders their diet more plentiful, but cheaper and wholesome.
>
> …When these gardens are of a proper extent, they fully occupy the leisure hours of the whole family, and thus not only tend to promote comfort, but to produce habits of attention and industry, and to cherish a taste for gardening in the young generation.

(The practice of cottars keeping a pig or cow is here observed, allowing the family to live off butcher meat for at least six months of the year.)

Two hundred years later, Neill's observations continue to resonate.

Prior to 1760, only six nurseries existed in Scotland. These totalled in aggregate some 40 of 50 Scotch acres. By 1818, the acreage of public nurseries had increased tenfold. Neill observed that the Perth Nursery was the third largest in Scotland at that time, containing 30 acres, whilst the nearby Dundee nursery consisted of only 12 acres. The Dickson nurseries had the greatest combined acreage in Scotland originating from one family: 'At Perth is to be found the best collection of hardy shrubs and fruit trees outside of Edinburgh'.

In terms of the development of nurseries post-1760, Neill considered tourism as an additional factor in increasing the taste for gardening and 'riches being now more equally distributed'.

Chapter 4

Perth Nursery – Grafting Innovation

Swede Turnips

> These Nurseries are also entitled to the distinguished honour of being one of the first, if not the very first, to distribute one of the most valuable roots the Scottish farmer has received from foreign countries – viz., the Swedish turnip, – a number of seeds having been sent to the firm in an envelope by the great botanist, Linnaeus, about the year 1772.
>
> Thomas Hunter, *Woods, Forests and Estates of Perthshire* (1883)

The turnip was an invaluable crop in the Agricultural Revolution, helping facilitate the Industrial Revolution by allowing efficient crop rotation to be established and thereby producing a more sustainable fodder source. Stewarts of Tayside, a matter of miles from the former Nursery, is now the largest producer of swedes in the country, turning over in excess of 45,000 tons per year – supplying 'neeps' to supermarkets across the length and breadth of the UK. (Commonly considered as yellow turnips, the swede is in fact a cross between a turnip and a cabbage.)

Scots Roses – A Fleeting Pleasure

Rosa spinosissima (R. pimpinellifolia) has numerous vernacular names such as the Burnet Rose, Scots Rose, and Scotch Rose.

Joseph Sabine FRS (1770-1837) was an English lawyer, naturalist, and horticultural journalist. Sabine provided the first account of the origin of the double 'Scotch Roses' in a lecture – 'Description and Account of the Varieties of Double Scotch Roses cultivated in the gardens of England' – presented in 1820 and later published in *Transactions of the Horticultural Society of London Journal, Volume 4* in 1822:

Description and Account of the
Varieties of Double Scotch Roses, cultivated in
the Gardens of England.
By Jospeh Sabine, Esq, FRS &c Secretary

Amongst the modern additions to the ornaments of our gardens, the varieties of Double Scotch Roses stand deservedly very high in estimation; their beauty is undisputed, and as they come into flower full three weeks before the general collection of garden Roses, they thus protract the period of our enjoyment of this delightful genus. On the British collector's notice they have an additional claim, being almost exclusively the produce of our own country; for of the many kinds that I have observed there are only three which can by any possibility be supposed to have originated out of Great Britain.

The Scotch Rose has been, and still is, sometimes called the Burnet Rose; it is the *Rosa spinosissima* of the English authors of authority who have written on the genus; they have united the *Rosa pimpinellifolia* and the *Rosa spinosissima* of Linneaus treating them as the same species, and not even separating them as varieties. Willdenow, however, in his Species Plantarum has adhered to Linneaus's distinction in which he seems to be supported by Jasquin in his *Fragmenta Botanica*; Miller also in his last edition of the Gardener's Dictionary kept them separate. The *Rosa pimpinellifolia* is the small wild (or what I would consider) the true Scotch Rose, with very small leaves, generally with smooth peduncles, and a flower (with very few, if any exceptions), more or less tinged with red; this is the *Rosa scotica*, or dwarf Burnet leaved Scotch Rose of Miller, and it is figured by Jacquin *as Rosa pimpinellifolia*. I do not mean to question the propriety of considering them as the same species, but they are assuredly, so different from each other, that they ought to have been treated as varieties and when all the plants usually called Scotch Roses are brought together, the *Rosa pimpinellifolia* above alluded to must be considered as the type of the species, for, if they have all been derived from one stock, I apprehend

Rosa spinosissima

Scots Rose – *Rosa spinosissima*
(*Royal Horticultural Society Lindley Collection*)

that was the original parent; for which reason if I were writing an account of the genus, on treating the particular species, I should certainly adopt *pimpinellifolia* as the specific name...

The Double Scotch Roses are more especially the object of attention with ornamental gardeners. They are nearly all strictly referable to the True Scotch Rose, or *Rosa pimpinellifolia* above mentioned, for the variations from the type, in foliage, and mode of growth, are very trifling in most of them; the chief difference between them is in the colours and the impletion of the flower...

The older books on gardening make no mention of the flower; even the last edition of Millers Garden Dictionary does not notice a double one. In the second edition of the *Hortus Kewensis*, though the list there given of cultivated Roses is large, not more than six varieties of the Scotch Rose are mentioned, only one of which is double, and that even is not properly a Scotch Rose; so that they are, in fact, altogether new subjects to the Writer.

The first appearance of the Double Scotch Roses was in the nursery of Messrs. Dickson and Brown (now Dickson and Turnbull) of Perth, between twenty and thirty years since. I am indebted to Mr Robert Brown, one of the partners of the firm at the above period, for the following account of their origin. In the year 1793, he and his brother transplanted some of the wild Scotch Roses from the Hill of Kinnoul, in the neighbourhood of Perth, into their nursery garden: one of these bore flowers slightly tinged with red, from which a plant was raised, whose flowers exhibited a monstrosity, appearing as if one or two flowers came from one bud, which was a little tinged with red: these produced seed, from whence some semi-double flowering plants were obtained; and by continuing a selection of seed, and thus raising new plants, they in 1802 and 1803 had eight good double varieties to dispose of; of these they subsequently increased the number, and from the stock in the Perth garden the nurseries both of Scotland and England were first supplied.

In Scotland, Mr Robert Austin, of Glasgow, a corresponding member of the Society (and of the firm of Austin and McAslan, nursery men in Glasgow), about 15 years since obtained the

Scots Rose – *Rosa spinosissima*
(*Royal Horticultural Society Lindley Collection*)

varieties from Perth, and has since cultivated them to a great extent, having now in his collection upwards of one hundred different new and undescribed sorts, some of which, perhaps when compared notice to the best now cultivated, may not be deserving of particular notice; but many are of such beauty, and so decidedly distinct, that, when made public they will greatly increase the catalogue of these ornamental plants.

In England, Mr William Malcolm, of Kensington, in the year 1805, purchased from the Perth nurseries six of their original sorts, and subsequently obtained the two others. They had been sold before that time to several noblemen and gentlemen, who were customers of Messrs Dickson and Brown. Messrs Lee and Kennedy of Hammersmith received the first of their stock from Mr Drummond Burrell, now Lord Gwydir, who brought them from Perth. The same kinds have since been obtained by Messrs Whitley, Brames and Milne, of Fulham. But, though the above three collections are by far the most complete of any, yet more or less of all the varieties are to be found in the nursery gardens outside London.

[Sabine proceeded to describe his own extensive collection of Scotch Roses and created a table of the Sections and Varieties.]

Section I: Double White Scotch Roses
1.   Small Double White
2.   Large Semi-Double White
3.   Large Double White
4.   Whitley's Double White

Section II: Double Yellow Scotch Roses
5.   Small Double Yellow
6.   Pale Double Yellow
7.   Large Double Yellow
8.   Globe Double Yellow

Musk and Velvet Roses
(*Royal Horticultural Society Lindley Collection*)

Section III: Double Blush Scotch Roses

9.   Princess Double Blush
10. Double Lady's Blush
11. Anderson's Double Lady's Blush
12. Dutch Double Blush
13. Double Provins Blush
14. Double Pink Blush
15. Double Rose Blush

Section IV: Double Red Scotch Roses

16.   True Double Red
17. Double Light Red
18. Double Dark Red

Section V: Double Marbled Scotch Roses

19. Double Light Marbled
20. Double Crimson Marbled
21. Double Dark Marbled

Section VI: Double Two Coloured Scotch Roses

22. Small Double Two Coloured
23. Large Double Two Coloured

Section VII: Double Dark Coloured Scotch Roses

24. Small Double Light Purple
25. Double Purple
26. Double Crimson

Robert Austin of Austin & McAslan in Glasgow had bred 210 named cultivars by 1827. It is not known how many cultivars Robert Brown raised himself after the first eight, but records from Woburn Abbey (of 1833) indicate that around 260 'Scots Roses' supplied by Dickson & Turnbull at Perth were grown in its *Rosarium Scoticum*. There were more named cultivars of Scots Roses in commerce in 1849 than all the main groups of roses (Albas, Gallicas, Centifolias etc) combined.

———

Rosa Centifolia.　　　　　*Rosier à cent feuilles.*

P. J. Redouté.　　　　　　　　Langlois.

Cabbage Rose – *Rosa centifolia* (*Royal Horticultural Society Lindley Collection*)

(*Elspeth Bruce*)

The 6th Duke of Bedford's head gardener at Woburn Abbey was a James Forbes, believed to have been born in St Madoes, Perthshire, in 1794. Another eminent head gardener, Forbes, inherited the renowned Humphrey Repton's landscape designs at Woburn. Forbes produced *Hortus Woburnensis*, a publication listing the 6,000 plant varieties within the gardens at Woburn Abbey. Within this tome Forbes wrote:

> Opposite to the Willow Garden is a large mass of Rock-work, lately formed, and planted with a choice collection of hardy alpine plants: upon the left of this, rises another bank of Rock-work, wherein exists a very complete *Rosarium Scoticum*, approached by a similar arched trellis, containing all the numerous varieties of the Scotch Rose, raised by Messrs. Dickson and Turnbull, whose Nursery at Perth, has been so long celebrated for this Rose, as well as for their very extensive collection of other ornamental plants.

1794

Rosa gallica
(Royal Horticultural Society Lindley Collection)

Just as it is impossible to assess the significant impact of the Perth Nursery in terms of the commercial growing of timber and the supply of ornamental plants and trees across the great estates of Britain, it is impossible to know how many cultivars of roses actually originated from the Perth Nursery. Undeniably, Brown's first double-coloured cultivars became the starting point for breeding a diverse range of new cultivars with at least 500 single, semi-double and fully double cultivars being available in a wide range of colours by 1840.

Whilst Brown is best known for his pioneering work with 'Scots Roses', he also raised new cultivars from seed trials, and around 1796, procured several seedlings of great beauty particularly the following:

Venus – a small double white, without prickles, probably the finest rose yet produced in Britain, raised from seed of the garden White Rose
Duchess – double blush, raised from Maiden Blush
Diana – double blush, raised from the same
Chance – double blush, raised from Double Damask
Victoria – dark and double, superior to the Tuscany
Mount Etna and *Mount Vesuvius* – raised from *Double Velvet (R. gallica)*
Parson – equal to Tuscany, raised from *Double Velvet (R. gallica)*
Ruby, Vagrant, Victoria, Fair Maid – raised from *Portland* or *Crimson Monthly (Rosa centifolia)*

Mr Brown possesses a new rose of considerable beauty, raised at New Scone, and which flowered in 1821. He calls it the Coronation Rose.

Patrick Neill FLS, *Journal of a Horticultural Tour Through Some Parts of Flanders, Holland and the North of France in the Autumn of 1817 – Transactions of the Caledonian Horticultural Society* (1823)

In the same journal, Neill described in considerable detail, a two-and-a-half month horticultural travel itinerary travelling from Leith to Flanders, Holland and France, via London and Canterbury. The membership of the deputation is unknown, although it is assumed that they would be the wealthier members of Society. Whilst in London they breakfasted at Lee and Kennedy's Vineyard Nursery in Hammersmith.

Rosa, Bella Donna
(Royal Horticultural Society Lindley Collection)

Born in Dumfries, Lewis Kennedy (*c.*1721–1782) served as gardener to Lord Wilmington at his estate in Chiswick, now West London. The estate's vineyard was notable for its production on an annual basis of 'a considerable quantity of Burgundy wine'.

Around 1745, Kennedy and James Lee of Selkirk (1715–1795) created a gardening partnership which went on to introduce commercially several important tropical and sub-tropical plants that could be grown and maintained in gardens or in greenhouses and hothouses. These included the China Rose (in 1787), *Fuchsia coccinea*, today *F. magellanica*, the fuchsia, the dahlia (in 1807), and the French rose as a standard (in 1818).

Lee was an experienced gardener who having undergone an apprenticeship at the Chelsea Physic Garden under Philip Miller went on to be gardener at Syon House for the 7th Duke of Somerset and gardener at Whitton Park for Lord Islay, later 3rd Duke of Argyll. It was whilst in the latter's service that Lee developed further as a gardener having access to the Duke's library, which was informed by the Duke's interest in gardening.

1 August 1817

We sailed from Leith to London on the 1st of August, in the *Czar*, one of those elegant and commodious vessels which ply between that port and London. We made prosperous voyage, having entered the Thames in little more than four days.

On 28 August, they visited a number of flower nurseries in Haarlem, including the bloemistry of Kreps and Company where they were shown the methods used in caring for hyancinth and tulip bulbs:

Being aware that Dutch nurserymen sometimes talk of furnishing three hundred varieties of Roses, we were desirous of viewing these in the nursery lines. We found the collections here to be very considerable; and Mr Kreps mentioned that he had procured all the known sorts cultivated in Holland, and many from England; but he candidly added that he could not, in fairness engage to furnish more than about one hundred distinct varieties. We may add, that, as far we could learn, the new varieties of native Scots roses, as well as garden roses, raised by

Rosa alba, var. rubicunda 'Celestial'
(*Royal Horticultural Society Lindley Collection*)

Messrs Brown at Perth and Mr Austin at Glasgow, excel, in delicacy of appearance, all the more recent productions of the florist in this department of the art, either at Haarlem or in any other part of the Low Countries.

These have been procured by sowing seeds of the semi-double varieties of the little Scots rose and seeds from the heps which frequently follow multiplicate flowers of Rosa alba, gallica and centifolia.

Sadly, 'Scots Roses' were a 'fleeting pleasure' both in terms of their relatively short flowering period and in terms of their popularity in the context of garden trends. The Scots Roses were most popular in the early nineteenth century, when their delightful strong fragrance, hardiness, and abundance of bloom proved them to be a very popular garden plant.

By 1821, the French rose breeder Calvert & Company in Rouen issued a catalogue featuring around 900 roses, of which 50 were Scots Roses. Calvert had imported some from Britain, some from French nurseries, and some they had propagated themselves. Around 12 cultivars of *R.spinomissima* were illustrated by Redouté and were described by Thory in *Les Roses* (1817-1824).

Scots Roses declined in popularity with the introduction of repeat flowering varieties and thereafter many cultivars became extinct in Britain and France. Very few Scots Roses are commercially available today, and it is said by horticultural conservationist Peter D A Boyd that many of them are misnamed. As of 2020, over 300 cultivars are conserved by Boyd in Shropshire as part of the Plant Heritage Conservation Initiative – this currently being the only collection of Scots Roses in Great Britain.

Trees

The Highland Society of Scotland, who patronise the improvements of the country with a liberality that excites emulation and industry among all ranks, have brought under the notice of Landed Gentlemen the importance of getting trees for planting raised from seeds collected from the soil and climate of which the trees are indigenous. Messrs Dickson and Turnbull, of Perth, according to the wishes of the Society early provided tree seeds

Douglas Fir, Dunkeld (*Roben Antoniewicz*)

and have at present a large assortment of pines especially the Pinus Sylvestris, or Scots Fir, from the native forests of Braemar and also Pinus Larix or Larch Fir, from Tyrol, of which these valuable varieties are both indigenous.

Morning Chronicle, 14 January 1830

Tree seeds were kiln-dried at a property in Canal Street, Perth, owned by Dickson & Turnbull, where a fire broke out in 1854, caused by an overheated kiln.

Through the exertions of the fire brigade, the adjoining properties, one of which was an extensive granary well stored with grain, and in great danger, were saved.

Northern Warder and General Advertiser for the Counties of Fife, Perth and Forfar, 16 February 1854

Loudon states that the first pinetum in Scotland was formed at Methven Castle, a matter of miles outside Perth, before 1830.

In the *Transactions of the Highland and Agriculture Society of Scotland* (31 January 1837), Thomas Bishop, Land Steward of Methven Castle, reported on the cultivation of certain species of pine. The report was transmitted to the Society in reference to an annual premium offered for the introduction of any new species of tree into forest, or ornamental plantations. Bishop described the new trees which he had successfully brought under cultivation on muir lands belonging to 'Robert Smyth, Esq, of Methven in the County of Perth'. He described planting up a pinetum in the spring of the years 1830 and 1831 in an area of muir ground 600 feet above sea level, comprising initially of 'larch and Scotch fir':

And here I began to form a Pinetum, being encouraged thereto by Messrs Dickson and Turnbull, nurserymen Perth, and others, who furnished me with species and varieties of the coniferous plants in their possession, at an under rate.

Bishop compared the characteristics of *Pinus cembra* (Swiss pine) in comparison to larch, finding the former to have greater durability.

After Bishop's initial successes with Swiss pine, he proceeded to purchase another 50 trees from the Perth Nursery; and continued to experiment with planting various trees in different conditions. The next new species that he wished to bring to the attention of the Society was *Pinus douglasii* (Douglas fir), which was introduced to Scotland from seeds collected by David Douglas on the northwest coast of America in 1827. Bishop wrote:

> During the few years that this species has been in the country,
> it has given great promise of becoming a valuable acquisition to
> its arboriculture, and in no situation more distinctly then on the
> muir ground above described.

His third recommendation was the *Pinus ponderosa* (ponderosa, bull or blackjack pine), first raised in Scotland in 1828 – again from seeds sent home by David Douglas from North America. Recommendation number four was *Pinus Caramanica* (Crimean pine), which was raised from seeds sent from France to Loudon, under the name of *Pinus resinosa*. Archibald Gorrie of Annat Garden had raised the plants for Bishop from this seed.

In 1838, Loudon published *The Suburban Gardener, and Villa Companion* and in the same year, the eight volume *Arboretum et Fruticetum Britannicum*. This publication made references to the Perth Nurseries, mentioning a number of maple varieties and other ornamentals such as *Magnolia tripeleta* (umbrella magnolia).

In a letter dated 16 May 1867, written by John Anderson (Turnbull's nephew) to Isaac Anderson Henry Esq, Edinburgh (an Edinburgh lawyer and horticulturalist), Anderson described how Brown & Turnbull were:

> [On] a Botanical excursion in the Highlands, being the time
> they discovered the *Menziesia caerulea*, and on their way back,
> as they were looking over a Glen about three miles West of
> Kenmore they noticed a branch of an Ash entirely yellow which
> they took away with them and budded it on the common Ash at
> Perth Nurseries. As far as I can ascertain from him none of the
> buds grew, and memory fails him in recollecting much about it,

only the fact that the operation communicated the disease or blotch to the stocks on which it had been budded, and it has been grown since that time, and annually grafted and catalogued under the name of Blotched Breadalbane Ash.

Anderson described further grafting experiments which were undertaken to provide new stock. Anderson Henry proceeded to write to Charles Darwin Esquire FRS in a letter dated 20 May 1867:

> *My dear Sir*
> *Happening to pass thro' Perth last week and having an hour to spare I visited the Nursery of which Mr Brown, who communicated to Dr Neill the extraordinary results of a graft I alluded to in a Paper I lately wrote on hybridisation, I thought I might learn some particulars of it from my friend Mr Turnbull of Belwood the Head of the existing firm now an old gentleman. I have not been disappointed (tho' I missed Mr Turnbull) as you will see by the enclosed letter I have from his nephew – of which make any use you please.*

In the absence of any plant catalogues, Thomas Hunter (1883) provides a comprehensive description of the trees being cultivated over 60 acres:

> We have said that the country at large has benefited by the establishment of these Nurseries, but we might even go further, and say that they have obtained a world-wide celebrity for their splendid collections of trees, including coniferae, ornamental shrubs, deciduous trees, rhododendrons, herbaceous and alpines, not to speak of 'florists' flowers, roses, stove, greenhouse, orchidaceous plants, &c...
> Great exertions are made during the season to get the stock into the necessary state for safe removal; and when it is mentioned that, at a recent stocktaking, it was found that there were over ten millions of trees of all kinds in the Nurseries, some idea may be formed of the magnitude of the operations entailed to bring them to a marketable condition. The immense number of seedling beds of ordinary kinds of fir is something astonishing, and no

one can visit the grounds without being interested in the enormous varied stock of all kinds, from the tiny seedling to the largest trees grown for single specimens for planting in parks and pleasure grounds. Although the proprietors have always aimed at enriching their collection by adding all the new introductions in every department, – especially if they are sufficiently hardy to stand the climate, – still the arboricultural experience of upwards of a century has not been thrown away. They have found that the larch fir seeds, of which large quantities are annually imported from the Tyrolese mountains do not thrive in these nurseries, and they consequently rely upon home-saved seed. This has also been their experience with native Scots fir. The imported seed is found to be useless, in consequence of its inability to stand the climate. They are, therefore, careful to sow nothing but home-grown seed.

In addition to the cultivation of enormous quantities of forest trees, such as larch, spruce, silver, and Scots firs, *Pinus laricio*, *P. austriaca*, *A. douglasii*, sycamores, beech, ash, oak, &c., thorns or quicks, for fences, are largely raised. The recent introduction of so many fine and hardy conifers from the Crimea, California, British Columbia, the Rocky Mountains, the Himalayas, China and Japan, has created a great demand for those best adapted for ornament and timber.

As the timber of many of those trees to which we have referred as having successfully stood the test of the extraordinary winters of 1878 and 1879 especially is of great value when grown in their native habitats, we will endeavour to give some idea of the estimation in which the most important of them are held, which may, in some measure, serve as a guide to planters.

The impetus which has lately been given to planting in the United Kingdom in consequence of the rapid disappearance of the forests in North America, Sweden and Norway, has led to special arrangements being made at the Perth Nurseries to supply the large and increasing demand for timber trees.

Of course, many years must elapse before their utility in this country can be finally demonstrated; but the experience already obtained at the Perth Nurseries and elsewhere is of the very greatest value, and will help to determine the future profitableness of the different varieties if extensively planted...

The superb Scarlet Hawthorn (*Crataegus oxyacantha punicea*) was likewise raised here.

As the object of this work is mainly to describe the condition of arboriculture in Perthshire, it would be somewhat foreign to our purpose to give a detailed account of those portions of the Nurseries occupied with fruit trees, roses, &c., not forgetting the extensive glass structures and their valuable contents. We may state, however, that we found throughout the whole of these departments the same care bestowed in the selection and cultivation as in the arboricultural section. Any one paying a visit to these extensive grounds will enjoy a treat that will long be remembered, and will meet with the utmost courtesy from the principals and officials connected with the establishment.

Hybrid Rhododendrons

In an article published in the *North British Agriculturist* (14 February 1849), the author remarked that one of the very best hybrid rhododendrons he had seen for early forcing had been raised from *R caucasicum* and *R arboreum* by Messrs Dickson & Turnbull of Perth.

Bearing Fruit

At the time of the publication of Neill's *On Gardens and Orchards of Scotland*, there were 'upwards of twenty orchards in the Carse of Gowrie, Seggieden, Glencairn, Glendoick, Kinfauns, Balthayock, Mill of Errol, Bogmill, Horn, Seaside, Fingask, Forgan, and Castle Huntly. These orchards, were let annually by roup [auction]'. Neill ranked the Carse of Gowrie orchards as second to those of Clydebank in terms of importance.

In a letter dated 3 December 1827 from a Patrick Matthew Esq, of Gourdiehill Orchard, Carse of Gowrie, to Neill in his capacity as Secretary of the Caledonian Horticultural Society, Matthew sent a selection of his 'specimen best keeping apples' in order that Neill could order grafts for the 'Experimental Garden'. The correspondence referred to the Caledonian Horticultural Society's ten acres at Inverleith which was later incorporated into the Botanics in 1864:

Grafts of any of them I shall afterwards send you as you may desire...After leaving you, I visited Clydesdale. The cheapness of the ground, and genialness of climate there, overbalance the expence of carriage to Edinburgh. Clydesdale is but an infant settlement of fruit trees compared with the Carse of Gowrie; and from its moderness has generally a more profitable selection, but not nearly the number of varieties.

He listed no less that 77 varieties of winter apples including:

56. *Maiden* tree an excellent bearer: fruit very acid, but one of the best kitchen apples that grows; it does not keep well. A seedling raised by Mr Brown of Perth.

The Stuart/Stewart of Annat Papers reveal that in the early 1800s, the Nursery was supplying apple trees, for example 'Apples on Paradise stocks, Ryder Apples, Standard Apples', as well as 'Pears, Dwarf Apples, Dwarf Pears', but the extent to which the Nursery contributed to the expansion of the Carse of Gowrie orchards remains unknown.

Early Adopters and Satisfying Crazes

The Perth Nursery was exceptional in terms of its scale and diversity of production, especially in timber. They were also noted for their ability to follow the demands of its customers, recognising crazes and responding accordingly. Such crazes included, 'The Gooseberry Craze', which swept the country in the early 1800s. 'Gooseberry Clubs' began forming, with gardeners cultivating over 2,500 varieties. In 1826, Perth Nursery grew a total of 192 varieties – with Archibald Turnbull detailing the Nursery's entire gooseberry holdings – by colour – to the *Transactions of the Caledonian Horticultural Society*:

You may make use of our list of gooseberries if you think proper to put it in your Society Memoirs. The inclosure where we have our gooseberries is nearly quarter of an acre. I am so much hurried, that I have no time at present to give a general description.

Reynolds' Golden Drop Gooseberry
(*Royal Horticultural Society Lindley Collection*)

Chapter 5

Clientele

P ERTH NURSERY'S customer base was substantial and embraced all parts of the country, primarily large estates and botanical gardens. By the mid-nineteenth century, it is highly likely that the surrounding suburban Victorian villa dwellers were also customers, as suggested by today's heavily-forested backdrop in and around suburban Perth. At that time, horticulture had become mainstream and offered fashions to follow as well as the more practical provision of produce for households. Horticultural societies were thriving, and Loudon's publications were quenching the thirst for horticultural knowledge and driving a period of unprecedented improvements. Not only was the Perth Nursery open for promenades as the forerunner of today's ever popular garden centres, but retail premises were also situated at 26 George Street, Perth, with evidence of this dating back to 1796 (*Statistical Account of Scotland, 1791-1799*) through to at least 1867 (Letter from John Anderson to Isaac Anderson Henry). These premises were leased from the Glover Incorporation of Perth.

The following provide an insight into the scale and diversity of the Perth Nursery enterprise at Kinnoull.

Annat Lodge and Annat Estate:
Lieutenant-General Robert Stuart (*also* Stewart)

The properties of Annat Lodge, Perth, and Annat Estate, Rait, were acquired by Lieutenant-General Robert Stuart on his return from service with the Bengal Army of the East India Company and the British Army in India. He had amassed a personal wealth of £55,000 (equivalent to almost £5 million today) during his colonial service. Stuart employed another prolific reformer, horticulturalist and agriculturalist, Archibald Gorrie, as his gardener and thereafter factor. Gorrie was born at Logiealmond, Perthshire, in 1777. As an adult, he took charge of the hothouse department in a Leith Walk nursery,

Portrait in Oils of Lieutenant-General Stuart
of Annat and Rait (1774-1820) by Baird, 1816
(*Courtesy of Culture Perth & Kinross*)

where he made the acquaintance of John Claudius Loudon, an intimacy that
was maintained until Loudon's death.

Gorrie contributed several papers to the *Gardener's Magazine* and to the
Memoirs of the Caledonian Horticultural Society. His obituary featured in
Volume 17 of the *Gardeners' Chronicle and New Horticulturalist* (1857):

To say that his mind was well stored with knowledge would be an inaccuracy of expression, for his knowledge was not laid up in a storehouse, but was so completely assimilated that his mental character grew, as it were, upon it, so that in communicating his thoughts to others he did not merely deal out the facts and ideas in the crude form in which they had been received: they became elaborated by the (perhaps to him insensible) operation of his powerfully original mind that it was impossible to draw the line of distinction between what was acquired knowledge and what was spontaneous thought...

We long regarded Mr Gorrie as one of the most intelligent writers on rural affairs and natural history which Scotland has produced; and Scotchmen will do well to cherish his memory of that one who in his quiet and unobtrusive life has done great and lasting good to their country, by his personal influence, by his genial writings, and by his improvements in the art of rural industry.

Conserved in the Perth & Kinross Archive, housed within the A K Bell Library, Perth, are the beautifully written annual notes of sales from Dickson & Brown to General Stuart. Their accounts read of a cornucopia of plants, trees and sundries – averaging an expenditure of approximately £4,000 per annum at today's valuation – recorded from 1802 to 1814. The following represents a small glimpse of the orders supplied by Dickson & Brown:

Trees
Including '500 Elms, 6,000 Thorns, 10 Spruce, 100 Silver Fir, 42 Walnuts, 200 Roans, Swedish Maples, Black Italian Poplars...'

Sundries
Including 'flowerpots, brickwall nails, garron nails, peach pruning knife, scythe stones, quarry picks, diamond pointed shovels, watering pans, watering squirt, buckskin gloves...'

Ornamental Plants

Including
'72 kinds of flower seeds'
(not described).

And:
White Jasmine
Ayrshire Roses
Virginia Creeper
Passion Flower
Clematis Flammula
Evergreen Honeysuckle
Rose Accacea
Rhododendron Ponticum
Azaleas
Budleya Globosa
Gum Cistus
Swedish Juniper
Lilacs
Coton Lavender
Butcher's Broom

Fruit and Vegetables

Numerous (See Appendix I)

Herbs

Curled Cress
Plain Cress
Coriander
Fennel
Pot Majoram
Dill
Garlick
Mustard
Rosemary Clary

Seeds

Perennial Ryegrass
White Clover
Red Clover
Yellow Clover

Exotics

Sand Leek Andromeda
Smilax
Gemma Tamarisk
Carolina Allspice
Colutea
Pomegranate

Annat Lodge and Annat Stable Offices still stand today, surrounded by beautifully landscaped mature gardens, punctuated by towering trees, doubtlessly sourced by Gorrie from the Perth Nursery. The Coach House was restored from a ruin in the 1990s into a very desirable property. The Lodge also remains as a single occupancy home, whilst the former orchard has since been developed for detached properties.

Annat Lodge on Kinnoull Hill, which Stuart purchased from the original builder is a house later associated with the artist Sir John Everett Millais and his wife Effie Gray; and later still with botanist and entomologist Francis Buchanan White.

Annat Coach House, Formerly the Stable Offices for Annat House
(*Elspeth Bruce*)

ART. III. *Plans and Description of Annat Lodge, Perth, the Property of Mrs. Stewart.* By ARCHIBALD GORRIE.

I FORWARD you the plans of a villa near Perth, the grounds of which I laid out in the year 1814. Annat Lodge stands on the highest part of the lands attached to it, on the east bank of the Tay; it overlooks the houses in Bridgend, the bridge, the Tay, and the city of Perth. Part of the lodge was built, and consequently the site fixed, before it was purchased by the late Lieutenant-General Robert Stewart.

In the plan of the estate (*fig.* 8.), which consists of about three acres, *a a* are the boundary fences; *b,* house; *c,* porter's lodge;

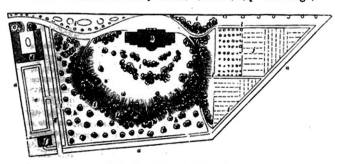

Fig. 8. *Map of the Grounds of Annat Lodge.*

d, offices; *e,* washing-green; *f,* a second porter's lodge; *g,* public service road; *h,* orchard; *i i,* flower-ground and shrubbery; *j,* kitchen-garden; *l,* walks.

Fig. 9. is an isometrical view of the house, showing the garden front, &c.

Fig. 10. represents the ground plan of the villa, in which *a* is the porch; *b,* vestibule; *c,* butler's room; *d,* housekeeper's room; *e,* kitchen; *f,* servants' hall; *g,* kitchen closet; *h,* scullery; *i,* back entrance to kitchen, &c.; *k,* principal stairs; *l,* china closet; *m,* dining-room; *n,* library; *o,* bedroom; *p,* bedroom closet; *q,*

Map of the Grounds of Annat Lodge

b, byre; *c,* stable; *d,* stairs to loft; *e,* larder over main cellar; *f,* coal-house; *g,* shed; *h,* wash-house; *i,* potato-house; *j,* coal-house; *k,* hen-house; *l,* piggery.

The villas in the neighbourhood of Perth display all sorts of styles, and many of them no style at all. Had your *Suburban Gardener* appeared twenty years sooner, and been read by the proprietors of these finely situated villas, a decided improvement must have been evident, both in the houses and in the grounds adjoining them. Since the publication of that work I have been continually pressing the subject of improvement on the attention of the owners.

Annat Cottage, Dec. 9. 1837.

Description of Annat Lodge

Fig. 9. *Isometrical View of Annat Lodge.*

butler's pantry; *r*, water-closet; *s*, fruit-room over cellars; *t*, well.   On the roof of the wing which contains the servants' hall

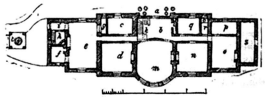

Fig. 10.   *Ground Plan of the principal Floor of Annat Lodge.*

and scullery there is a cistern for the supply of water.

Fig. 11. is a plan of the offices, in which *a* is the porter's lodge;

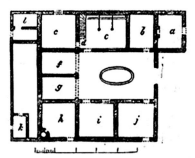

Fig. 11.   *Ground Plan of the Stable Offices of Annat Lodge.*

Plans and Description of Annat Lodge

Atholl Estates:
John Murray, 4th Duke of Atholl, KT, PC, FRS

The amount of forestry planting undertaken by the great landed proprietors between 1775 and 1850 was exceptional. Said to be the greatest tree planter of all was the 4th Duke of Atholl, also known as the 'Planting Duke' or 'Planter John'. Between 1774 and 1826, he is said to have overseen the planting of over 14 million larches and over 13 million trees of other kinds.

'Planting should be carried on for Beauty, Effect and Profit.'
'Planter John', 4th Duke of Atholl

The Duke's ambition was to supply the entire British Navy with trees from his estate. Unfortunately, the development of the steel industry was to scupper such plans.

The Factor's Accounts for the period 1867 to 1920 contain numerous purchases made by the Atholl Estates from Dickson's for 'seeds', 'implements', 'plants' etc, but whether for Blair Castle or Dunkeld, is unknown.

Diana's Grove, Blair Castle
(*Courtesy of Atholl Estates*)

Fingask Castle:
Sir Patrick Budge Murray Threipland, 4th Bart

The Fingask landscape (Perthshire) is included in the inventory of gardens and designed landscapes as one of the oldest gardens in Scotland. The terraces and woodland around the Category A-listed castle date from the late seventeenth century, while the nineteenth century topiary and woodland gardens are described as having 'high artistic value'. Fingask Castle archives include information as to vegetable and fruit production and the purchase of seeds and other items from Dickson & Brown. A walled garden remains in use at Fingask today, with the castle and grounds offering a popular venue for weddings and celebrations.

Balloch and Taymouth Castles:
3rd and 4th Earls of Breadalbane

Around 1550, Balloch Castle, as Taymouth was then called, was owned by Sir Duncan Campbell, 1st Baronet. According to *Burke's Peerage*, he is reputed to have been the 'first Highland laird to turn [his] attention to rural improvements and plant trees and he forced his tenants to do so'. John Campbell, the 2nd Earl of Breadalbane, who succeeded in 1717, initiated alterations to the castle and laying out of the grounds. He died in 1752, aged 90. The 3rd Earl is said to have laid out the first design in 1720 and laid out a further landscape in 1750 – the subject of the painting 'Birdseye view of Taymouth Castle', attributed to James Norie.

Taymouth Castle archives (Breadalbane Papers), conserved in the National Records of Scotland, indicate the level of improvement being undertaken by the 3rd and 4th Earls. Between 1800 and 1834, the 4th Earl transformed the castle using various architects including William Atkinson who was noted for his interest in gardening. (In 1831, John Campbell, 4th Earl of Breadalbane was created the 1st Marquis of Breadalbane; he died three years later.)

The Breadalbane Papers also reveal petitions raised, mainly from tenants, and are addressed either to the Earls themselves or to their factors. (A petition was the recognised manner of drawing a felt injustice to the attention of the Earl or his commissioners.) During the latter part of the eighteenth century and the early part of the nineteenth, most of the petitions are in the same hand. Tenants were rarely able or willing to draw up their own; some are signed, either with a signature or mark, but many are not. Breadalbane tenants showed a marked indifference to the march of agricultural science, wishing for things to remain as they were. The 3rd and 4th Earls were improvers, building dykes, farm steadings, roads and bridges, first wooden ones and then stone, mills of various kind and then latterly planned villages; and introduced tradesmen and manufacturing. Disputes between tenants and landlords were commonplace as they continue to be, tenants now being assisted to a degree by the Agricultural Holdings (Scotland) Act 2003 and the Scottish Tenant Farmers Association seeking to support tenant farmers.

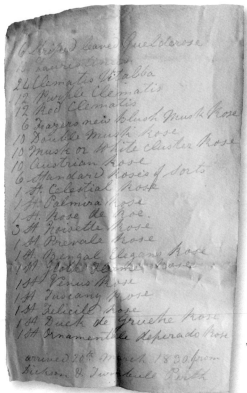

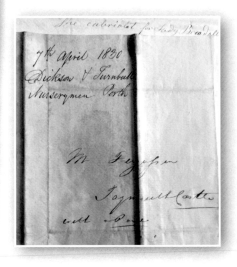

Taymouth Castle Order for Roses, 1830
(*National Archives of Scotland*)

The archives reveal continuous estate improvements as well as interesting insights into the sometimes strained relationships between master and servants. In regards to Dickson & Brown, the following is recorded:

1771   (GD 112/74/353) – amounts due for 20,000
seedling oaks and 50 horse chestnuts.

1772   an account due per order of Mr Alexander
Thomson for various vegetable seeds, implements
and flower pots.

1773   an account due for vegetable seeds, flowers and
sundries from Dickson & Brown to the Earl of Breadalbane.

1778-9   an account due for 10,000 each of seedling oaks
and larix, and 100 large service (trees).

1780-1   an account due for garden seeds and sundries,
including Spanish morels [peas], salsify, skerret,
cardoons, herbs, amaranthus coxscomb, 12
hotbed mats, Pomfret cabbage, 20 best kinds of
flower seeds gratis, kidney potatoes, asparagus,
and fruit trees.

1787-8   an account due for vegetable seeds, seedling trees
and garden sundries.

1788   copy account due for seeds and garden sundries
including ten kinds of bean and nine kinds of
peas, and canteloup melons, walnuts, sweet
chestnuts, Weymouth pines, black American
spruce and some mushroom spawn.

1797-9   an account due for seeds and shrubs, including
narrow leaved myrtles, geraniums, cineraria
lanata, Scots laburnum, Patagonian cucumber,
mushroom spawn, 36 poplars. Paid 14 January 1808.

1800-1   an account due for thousands of seedling larch
and Scots firs.

1817   an account due for seeds for Taymouth in 1815;
and an account for seeds and sundries supplied
for garden at Achmore, 1815.

1830   numerous purchases by Lady Breadalbane
'Tuscany' and 'Félecité' roses and various types
of clematis. Other trees and Portugese laurels
were also purchased.

1831   Lady Breadalbane's order for carnations.

In his *The Encyclopaedia of Gardening* (1822), Loudon described Taymouth as 'the most magnificent residence in the County' (Perthshire):

> The house a spacious gothic mansion erected at different times placed on a lawn about a mile in breadth between two mountains which open to Loch Tay on one side and the Tay River passes within two furlongs of the house. The mountains, lawn and the banks of the waters are richly clothed with wood, through which are laid magnificent walks. Of the trees the limes and larches have attained a great size and there is an avenue of the former 450 yards in length scarcely equalled anywhere.

The grounds of Taymouth Castle underwent considerable change over a number of years, especially in the eighteenth and nineteenth centuries. Judging by the plant orders placed with the Perth Nursery and the passion for landscaping and gardening shown by its owners, it would have met with Queen Victoria's and Prince Albert's approval when they paid an official visit to Taymouth in 1842. The Queen returned in 1866 for a private visit. The contents of the Breadalbane Papers appear to be without any censorship and provide a truly evocative read.

Taymouth Castle (View of Taymouth Castle from the South)
James Norie and Jan Griffier II.
(*Photography Copyright National Galleries of Scotland, photography by Antonia Reeve*)

The Royal Botanic Gardens, Kew

Purchases by the Royal Botanic Gardens are evidenced in Kew's archives but are insufficiently described to identify the actual existing vestiges today, being listed only as '*Coniferae*, Shrubs and rhododendrons'. Between 1868 and 1878, over a thousand trees and shrubs were purchased from Dickson & Turnbull by the Royal Botanic Gardens. The Royal Botanic Gardens' goods inward records reveal:

1849	a collection of Hybrid Rhododendrons consisting of 100 plants six inches high.
1872	1,013 Coniferae.
1873	76 Coniferae.
1874	24 Coniferae.
1876	110 Trees and Shrubs.
1878	305 coniferous and other shrubs.

The *Gardeners' Chronicle and Agricultural Gazette* of 22 July 1871 described, in an article, improvements and developments at Kew Gardens, by Sir Joseph Hooker. The following provides a complimentary testimonial to the Perth Nursery from an eminent horticulturalist:

> Between 1840 and 1865 many efforts were made by my predecessor to keep up the sylvan scenery of the pleasure grounds by planting conifers amongst the old trees in every available open space especially Deodars, Cedars, Scotch, Douglas, Austrian, Corsican and Weymouth Pines. Pinus longifolia, Smithiana, Anne spruce is of various sorts besides forest trees innumerable but his permission could not be obtained either to make sufficient clearances or to disturb the roots of the old trees by trenching in the ground these plantations have utterly failed.
>
> The Lake in the Pleasure Grounds which was half finished in 1869 has since been completed and the whole of the ground on the south side of it cleared, covered with good soil and prepared for the formation of the new Pinetum the planting of which will have begun forthwith. In reference to this Pinetum I have to state that as no complete public arranged unnamed collections

of hardy Conifers exists in England, the establishment of such an one at Kew is looked forward to with much interest by both collectors and nurserymen. Of the latter, two of the most eminent, Messrs Lawson of Edinburgh, and Dickson and Turnbull of Perth have presented to the Royal gardens specimens of every species and variety that was to be found in duplicate and their extensive collections. Mr McNab of the Edinburgh Botanic Gardens has also sent many valuable plants for this Department.

Purchases were also made by Dickson & Turnbull from Kew in 1882, which were more detailed. The orders consisted of a diverse range of plants, including viburnums, roses, streptocarpus, berberis, pines, gladiolus, and rhododendrons.

Walter Francis Montague Douglas Scott,
5th Duke of Buccleuch, 7th Duke of Queensberry, KG, PC, FRS, FRSA

The Duke was a great Scottish land magnate, owning vast estates in Britain, including Drumlanrig, Dalkeith, and Beaulieu Abbey. The Dalkeith Chamberlain's vouchers (1850), stored in the National Records of Scotland, reads as a horticultural 'Who's Who':

CLAPTON NURSERY, London, for camellias, Otaheite oranges, Kitley's goliath strawberry (established by John Bain Mackay (1795-1888), *b.* Echt, Aberdeenshire).

HUGH RONALDS & SONS, Brentford, for exotics etc. (established by Hugh Ronalds (1726-1788), *b.* Moidart, Inverness-shire).

ROBERT GLENDINNING, Chiswick Nursery (established by Scot, *James Scott*, a pineapple specialist, in 1740. Glendinning was born in Lanark in 1805 and took over the Nursery in 1843. He had an excellent reputation for growing conifers and fruit trees; was awarded numerous RHS awards; and later became one of the Chiswick Improvement Commissioners).

Clematis purpurea
(*Royal Horticultural Society Lindley Collection*)

DICKSON & TURNBULL, Perth.

JAMES VEITCH & SON, Exeter (established by John Veitch (1752-1839), *b.* Ancrum, Jedburgh, apprenticed at Hassendean-burn where he worked with James Lees, and later completed his apprenticeship with Lees of Hammersmith).

The chapter has offered but a small number of notable clients of the esteemed Nursery – exactly how many clients existed, will in all likelihood never be known.

_____

Chapter 6

The Spirit of Exploration

Do not follow where the path may lead. Go instead where there is no path and leave a trail.

We have now no means to ascertain when he formed the idea of becoming a botanical traveller, but we are inclined to think it may be ascribed to his intercourse with Messrs Robert and James Brown of the Perth Nursery, both of whom were good British botanists, and so fond of study as annually devote a part of their time to botanising the Highlands. Hence their excursions were often the topic of conversation, and, from hearing them recount their adventures, and describe the romantic scenery of the places they had visited in search of plants, he probably formed the resolution of imitating their example.

John Claudius Loudon, 'Biographical Notice of the late Mr David Douglas', *Gardener's Magazine* (1836)

DAVID DOUGLAS was born in 1799 in the village of Scone. At a young age, he would very probably have been uprooted from his home by the 3rd Earl of Mansfield as a result of Loudon's plans to redesign the grounds at Scone. From the age of seven to eleven, Douglas attended Kinnoull School, a matter of a minute's walk from the Perth Nursery. It is said that he studied flora and fauna on Kinnoull Hill and thus the reason he was often late for school. It is also highly likely that he would also have explored the grounds of the Perth Nursery, with its glasshouses, walled gardens, potting sheds, and follies; mentored by Robert Brown.

In 1823, Sir William Hooker recommended Douglas to the then Horticultural Society of London, who were looking for a suitable plant collector to send to North America. Douglas left his homeland and spent the latter

Perth From Kinnoull Hill (*Courtesy of Culture Perth & Kinross*)

half of the year based in Philadelphia, making contact with other botanists and collecting plants. On the welcome receipt of the notes and seeds which Douglas sent back to the Horticultural Society, the Hudson's Bay Company then offered to sponsor a Horticultural Society collector to explore the Columbia River in the Pacific Northwest. Consequently, the next expedition (1824) was to the west coast of North America, reaching the Columbia River where Douglas began collecting plants including the Douglas fir (given the botanical name *Pseudotsuga menziesii* – after Archibald Menzies, a fellow Scot and native of Perthshire). On two expeditions to the west coast of Oregon and up the Columbia River and its tributaries, Douglas collected over 200 species of plants including lupins, phlox, penstemon, sunflowers, gaillardia, and mahonia, subsequently transforming many British gardens and greenspaces.

Black Rock Viewpoint Overlooking Kenmore. *Larix decidua* (European Larch) and *Pinus sylvestris* (Scots Pine) in the Foreground
(*Copyright Perthshire Picture Agency*)

Douglas wrote back to his mentor, Sir William Hooker: 'You will begin to think that I manufacture pines at my pleasure', describing the wealth of trees he came across. He introduced Douglas fir, Sitka spruce, noble fir, grand fir, Monterey pine, and many others to Britain where they were received with great excitement. Douglas also fed the growing enthusiasm for landscaping with Sitka spruce, which continues to have an impact on commercial forestry in Britain to this day. Douglas was still plant hunting when he died mysteriously in July 1834 in Hawaii.

It was also in 1834 that Robert Brown, then aged 67, embarked upon a self-funded journey of horticultural exploration to North America and Canada, accompanied by the young botanist and horticulturalist, James McNab, son of William McNab, Superintendent of the Royal Botanic Garden Edinburgh. McNab maintained a journal during their seven-month tour (May-December 1834). It is a compelling account of their travels and challenges with references to critiques of American plant nurseries, familiar names with Scottish origins, such as 'Airleywright' and tales of the hospitality they enjoyed or otherwise:

14 July 1834.

At 9 o'clock am, we left Troy by the Whitehall Coach for Stillwater 16 miles further up the River in order to spend a few days at Airleywright the property of Mr Wylie formerly of Perth. A few miles after leaving we crossed the Hudson river and its tributary by means of a bridge and proceeded along its left bank, and were much pleased with this part of the country as a farming district, being undulated and well cleared. The Hudson River and its tributary streams affording the inhabitants an easy source of getting their goods to market.

Graham Hardy, 'To get to those plants likely to prove interesting at Edinburgh: 'Robert Brown of Perth and James McNab's North American tour of 1834', *Sibbaldia: The Journal of Botanic Garden Horticulture*, Number 9 (2011)

McNab and Brown made the return to Scotland from Philadelphia with over 200 plants, including varieties of roses. McNab described three new species and three new varieties in his paper published in the *Edinburgh New Philosophical Journal*:

> *Patrinia longifolia* (now *Valeriana edulis*), *Liatris borealis*, *Liatris stricta*, *Lobelia cardinalis* var. *alba*, *Gentiana angustifolia* var. *nana* (now *G. autumnalis*) and *Gentiana barbata* var. *browniana* (now *Gentianopsis virgata*). [The types of these taxa are now in the herbarium at Glasnevin, Dublin.]

Poinsettia – *Euphorbia pulcherrima*
(Royal Horticultural Society Lindley Collection)

One plant, now ubiquitous in nurseries and shops in the UK, particularly in the winter months, which McNab introduced through contact with the nurserymen he visited on the tour is the poinsettia, *Euphorbia pulcherrima* – an attribution which requires recognition, especially given its widespread popularity and contribution to commercial horticulture today.

After three years back in Scotland, Brown took the decision to emigrate to Philadelphia, joining other expatriate Scots, including the nurseryman and author Robert Buist, who had trained at the Edinburgh Botanic Gardens and emigrated to America in 1828. Buist took employment with Henry Pratt who owned Lemon Hill, probably one of the finest gardens in the US at the time. He formed a partnership with Thomas Hibbert in 1830 in a floristry business in Philadelphia. They imported rare plants and flowers, especially the rose. After Hibbert's death, Buist began a seed business, along with the nursery and greenhouse business, the Robert Buist Company. Buist was known for his roses and verbena and was credited with introducing the poinsettia to the US. Brown died in Philadelphia in September of 1845, leaving no known descendants.

An unlabelled handwritten note was found by Peter D A Boyd, (an authority on Scots Roses), within the archives of the Royal Botanical Garden Edinburgh which is believed to be a copy of the epitaph to Robert Brown:

> Sacred to the memory of Robert Brown, native of Perthshire, Scotland, who departed this life on the 21st of September 1845, in the 78th year of his age. Mr Brown was long and favourably known in the Horticultural and Agricultural World. To the first he contributed the double-flowering Scots Rose, and the Scarlet flowering Thorn. To the second the Swedish turnip, which he was first to introduce into Britain, whence it rapidly spread into other countries. The Scientific Botanist is indebted to him as the original discoverer of the Menziesia (now called Phyllodoce) caerulea, and the other rare natives of the Scottish Alps. In his 70th year he arrived in this Country, locating himself in the vicinity of Philadelphia, where the affability of his manners, and the fervency and sincerity of his affection secured him many friends, who fully appreciated his scientific attainments, and who have indulged their predilections by erecting this humble monument to his memory.

Another horticultural 'Son of Perthshire' is John Jeffrey, born in 1828 at Forneth, in the Parish of Clunie. The founders of the Oregon Association subscribed to send Jeffrey on a plant-collecting expedition, effectively to pick up where the ill- fated David Douglas had left off:

> The first expedition for the further exploration of Vancouver's Island and British Columbia was formed in 1849, under the patronage of the Hon Lord Murray; Sir William Gibson-Craig, Bart; Sir David Dundas, Bart; Professor Balfour; Messrs George Patton, David Smith, Charles Lawson, William T Thomson, Humphrey Graham, Archibald Turnbull, Isaac Anderson-Henry, Andrew Murray, Charles McIntosh and J McNab.
>
> John McNab, 'Discoveries of Mr John Jeffreys and Mr Robert Brown, collectors to the botanical expeditions to British Columbia between the years 1850 and 1866',
>
> Transactions of the Botanical Society, Volume 11 (1872).

John McNab was Curator of the Royal Botanical Garden Edinburgh at the time of his account and a subscriber to the Oregon Association. The Robert Brown referred to by McNab is yet another of that ilk. McNab strongly recommended Jeffrey for the expedition:

> After the arrangements for the proposed Expedition had been completed and subscribers obtained, I was asked to recommend a person fit to carry out the objects of the association. I at once named Mr John Jeffrey who had been for several years employed in the Edinburgh Botanic Garden. He was an energetic, painstaking young man, and capable of enduring hardship. By his volunteering to ascend a high tree to remove a branch which had been broken by the wind, a task few men would have undertaken I was first induced to take an interest in him. He had a great desire to obtain a knowledge of trees shrubs and herbaceous plants collecting drawing and naming all that came within his reach both British and exotic. After having been furnished with the requisite instruments and instructed in their use Jeffrey left Scotland in March 1850.

Jeffrey arrived at Hudson Bay in August 1850 and travelled more than 1,200 miles overland to reach the Columbia River, thereafter spending the next four years exploring Washington, Oregon, and California, sending his specimens back to Scotland. At the time, Jeffrey was criticised for poor results, a situation which McNab further defended, attributing the alleged poor results to a number of issues including lack of knowledge of cultivating the seeds, confused nomenclature etc, effectively attempting to clear Jeffrey's name and perhaps McNab's own reputation:

> I have prepared this notice of Jeffrey's collections from having observed remarks at various periods, and the Gardener's Chronicle which convinced me that his discoveries are not fully known to the public. From my knowledge of his doings, I am able to state that he introduced nearly as many new conifers as the late David Douglas while travelling in British Columbia between 1825 and 1833.
>
> Douglas's collections were sent direct to the Horticultural Society of London, examined and named from dried specimens and cones, and the seeds distributed to the members. Jeffery's collections were sent to the British Columbia Botanical Association at Edinburgh and distributed to the subscribers. When the pockets were large all received in proportion to the number of their shares but when small they were distributed to the shareholders by rotation. By this means some of the rarer species no doubt fell into the hands of distant parties who did not report upon them. I think I may safely affirm however that all the conifers are now known...
>
> As most of the circumstances connected with these Expeditions are known to me from the commencement I am anxious to give as far as I can a detailed notice of all the conifers discovered and introduced by the collectors to the Association. From the time Jeffrey went out until his death he sent home dried specimens of not less than 1,700 species consisting of trees shrubs and herbaceous plants all consecutively numbered with references to his diary which unfortunately never came to hand authentic particulars of his death never reached this country. It appears that he was killed when trading with the Indians.

McNab proceeded to name a number of conifers introduced by Jeffrey, including *Abies pattoniana* which was previously unknown in Britain, the finest specimens of which were known to McNab at Glen Almond and the Cairnies, where they were attended to by the late proprietor whose name was commemorated in the species. (Presumably this report referred to Messrs George Patton, a relative of the Dicksons.) Among Jeffrey's best known discoveries are Jeffrey pine (*Pinus jeffreyi*), tall mountain shooting star (*Dodecatheon jeffreyi*), and foxtail pine (*Pinus balfouriana*).

Chapter 7

The Germination of the 'County Show'

A RCHIBALD TURNBULL contributed to improvements quite literally in 'other fields'. In 1833, Dickson & Turnbull hosted 'The Exhibition of Agricultural Productions, New Implements at the Premises of Dickson & Turnbull, Nursery and Seedsmen, Perth', at the Nursery which was open to the public from November 1833 to April 1834:

> The eminent success which attended the Stirling Agricultural Exhibition, by Messrs Drummond, has led to a similar exhibition in Perth, a county which has long been distinguished both for horticultural and agricultural eminence. It is gratifying to find, from the preface of this, we should have been fully convinced of it by the manner which both farmers and horticulturalists have come forward with samples of their various productions.
>
> Among the exhibitors, we find our valued correspondent Mr Gorrie, and also Mr Bishop of Methven Castle and Mr Young of Pitfour stand pre eminent. Among the articles exhibited were, wheats in straw, twenty nine specimens, including half as many varieties; some from Tuscany, Tangier, Nepal, South America &. Of barley in straw, there were twenty five specimens; including one from Morocco, which yields 20 sacks per acre. Of barley in sample, there were seventeen specimens, including one from Tangier, and one from China. Peas of two sorts; including the Napoleon pea from Mr Gorrie... Tares, two sorts. Turnips, sixty five specimens. Carrots, sixteen sorts from gardens, and two from fields; the heaviest of these was an Altrincham weighing 3¼ lbs. Potatoes, twenty six sorts including an early white variety from Aberdeen, which produced, on a small spot, at the rate of 222 bolls of 560 lbs each per Scotch acre.

We may also mention 'a permanent white variety, obtained by cutting out and planting the variegated part taken from the red potato, performed by a female horticulturalist, from Mr H Bishop, New Scone.

. . . In thanking the female horticulturalist for the result of her ingenious experiment, Messrs Dickson and Turnbull gallantly observe 'If ladies would only take the lead, the young farmers would soon follow them in studying the science of vegetable physiology. We entirely concur in this opinion; and only wish that, in addition to vegetable physiology, ladies would study the natural system of botanical classification; direct their attention to trees and shrubs, so as to introduce a greater variety in our shrubberies and plantations; and not to forget altogether landscape gardening and domestic architecture and furniture'. . .

Among the fruits exhibited were some hundreds of varieties of pears and apples, including many of the new sorts introduced from France and the Netherlands; also a small branch, about seven inches long, containing about 100 cherries, which grew in Perth New Row Green, Perth, twenty two years ago, preserved in a glass (in spirits we presume.)

Numerous pines and firs in pots were exhibited from Dickson & Turnbull's own nursery. Models of various kinds were also exhibited including three ploughs made by ploughmen, as were portraits of prize cattle, cheese presses and models of drains, cheeses and books.

Loudon observed how easy it would be for 'seedsmen to get up similar exhibitions in every county town':

[The] object is less to get new and strange articles, than to assemble together such things as are already in the county or district, in order to facilitate comparison, and equalise the knowledge of the existence of such things. This alone would lead to the examination of the exhibitions of adjacent counties and to the introduction of articles from them; and thus improvements of every kind would surely and rapidly be propagated throughout the country. We would direct the attention of agriculturalists

everywhere, and more especially in England, to the great
number of varieties of wheat, barley, and oats exhibited at the
agricultural museums of Perth, Stirling and Edinburgh and to
the superior excellence of some of them.

John Claudius Loudon,
'Report of the Agricultural Exhibition at Dickson and Turnbull's, Perth',
*The Gardener's Magazine and Register of
Domestic & Rural Improvement*, Volume 10 (1834)

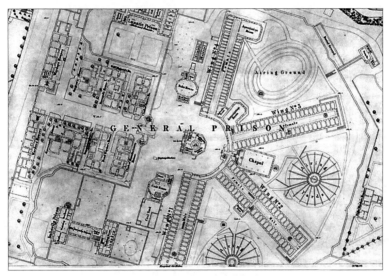

Map of Perth Prison in 1860

Two years later, the Highland Agricultural Society held its annual show on
7 October 1836 at the 'The Depot' in Perth, which had been built in 1812 to
house French prisoners captured during the Napoleonic Wars (today part of
HM Prison Perth). Many of the original buildings remain. Volume Seven of
the *Quarterly Journal of Agriculture* describes 'The Depot':

> There is perhaps no place so well adapted for the great shows
> of the Highland Society as The Depot at Perth. Independently of
> its centrical situation to extensive breeding districts, it is
> peculiarly suited, by its various compartments, to accommodate

in comfort and safety a very large number of every species of stock. The building was erected by Government for the confinement of French prisoners, of whom not fewer than 8,000 were accommodated in it. The range of buildings consist of five wards, the subdivision walls between which converge towards a common centre, which is occupied by a hexagonal battery and tower, commanding a fine bird's eye view of the whole buildings. The stalls and pens for the livestock were ranged along the fences of the wards and subdivision walls. One ward contained the extra oxen, cows, heifers and the sweepstakes, which were not entered for a competition; the next ward contained bulls and heifers; a third oxen and cows; a fourth sheep; and the fifth, horses. The pigs were arranged along the outer wall of the tower, and the roots and seeds in the verandas of the large guard houses at the entrance gate; and the implements along the street from the entrance to the inner gate. A large gallery for ladies was erected in front of the battery wall, commanding a complete view of all the wards. The platform for the exhibition of prize stock was placed in front of the gallery, and the gentlemen of the committee and judges were accommodated in various convenient and comfortable apartments. It will be seen, even from this imperfect sketch of the depot, that superior accommodation could not have been obtained. The noblemen and gentlemen forming the various committees were entertained on Thursday morning, by the Lord Provost and Magistrates of Perth, at an elegant, *dejeune a la fourchette* in the Assembly Rooms (a light lunch at the Salutation Hotel.) The journal lists the various classes, Dickson and Turnbull having entered under the Roots and Seeds category.

The next exhibition to be noticed is a collection by Messrs Dickson and Turnbull of Perth. This besides containing numerous examples of the potato and other useful plants, contained a very extensive collection of coniferous trees. The Committee recommended that an honorary premium be awarded to Messrs Dickson and Turnbull.

Archibald Turnbull had joined the Highland and Agricultural Society in 1826. Despite the constant drive for agricultural improvement, in 1846 potato crops were devastated in Scotland and destitution boards were set up to raise money for people in the Highlands and Islands who were faced with starvation.

Glasshouse Built at The Highland and Agricultural Show,
South Inch, Perth, 1879
(*Magnus Jackson, Perth Museum & Art Gallery, Culture Perth & Kinross, MJ784*)

The Perthshire Agricultural Society was formed later, in 1866; their annual show moving to the South Inch, Perth, from Forteviot in 1905. Both the Royal Highland Show and Perth Show continue to the present day.

Erica thunbergeii, Erica glauca, Erica pinfolia discolor
(Royal Horticultural Society Lindley Collection)

Chapter 8

Enterprise in Art

T
HE PAINTER John Everett Millais was born in Southampton in 1829, and considered a child prodigy, winning a silver medal at the Society of Arts at the age of nine. In 1840, he was admitted to the Royal Academy School as their youngest ever student.

In 1855, amidst scandal, he married Effie Gray, the art critic John Ruskin's former wife, with whom he had fallen in love while he was holidaying with the Ruskins in Scotland. Millais and Gray were married in the drawing room at Bowerswell, Perth, Effie's family home, adjacent to the Perth Nursery and Annat Lodge. After their honeymoon in Scotland, the couple took up a tenancy at Annat Lodge, Lieutenant-General Robert Stuart's former home, with its large garden and separate carriage house. Millais used the carriage house for painting. The Lodge is located on a rise just above Bowerswell. Annat was Gray and Millais' primary residence for just over two years.

In 1855/6, Millais painted *Autumn Leaves*, described as his 'most lovely picture' in the cedared environs of Annat Lodge, looking westward down the hill to Bowerswell and beyond towards the peak of Ben Vorlich near the centre of the horizon. Apple trees feature in the garden on the right, tall poplars stretch into the illuminated dusk sky and the bell tower of St John's Kirk is dimly visible. It must be assumed that these trees originated from the Perth Nursery, given Archibald Gorrie's connections with Annat and the relevant timeframes.

Millais's next two paintings which featured the Annat Lodge locality were *Spring* and *The Vale of Rest* which are thematically linked and almost identical in size. *Spring* was painted in the mid 1850s while Millais was with his family at Bowerswell. It is an image of eight picnicking girls arrayed against the low wall of an apple orchard. Would this be the orchard at Annat Lodge, stocked from Perth Nursery? It is very likely.

The scythe in the painting links this to *The Vale of Rest* as a symbol of mortality. In his painting *The Vale of Rest*, Millais was purportedly inspired by yet another sunset over Perth. The view blends the garden wall at Bowerswell with oaks and poplars and the old Kinnoull Parish Churchyard.

'Autumn Leaves', John Everett Millais, 1855-6
(*Image Courtesy of Manchester Art Gallery*)

'The Vale of Rest', John Everett Millais, 1858-9
(*Image Courtesy of Manchester Art Gallery*)

Millais was later criticised for his entrepreneurial approach to his painting *Bubbles*, originally titled *A Child's World*, painted in 1886 – becoming celebrated and renowned when it was used over many generations in advertisements for Pears soap. During Millais's lifetime, the painting led to widespread debate about the relationship between art and advertising. Millais became one of the wealthiest artists of his day and in 1896 was elected President of the Royal Academy. A statue of Millais stands at the entrance of Tate Britain, London. The large west window of Kinnoull Church was presented by George Gray of Bowerswell in 1870. Its design, painted on glass panels, depicts 14 parables, based on drawings made famous by Millais. Effie Gray is interred at Kinnoull Graveyard

Effie Gray (Lady Millais) by Thomas Richmond, 1851
(*Courtesy of Universal Art Archive/Alamy Stock Photo*)

Chapter 9

What became of the Perth Nursery?

ANDERSON & TURNER dissolved Dickson & Turnbull in 1881 and sold the business in 1883 to J McLeod, a nurseryman in Crieff. John A Anderson died in Perth on 6 December 1891. The company continued to trade under different ownership, eventually winding up any horticultural trading with the closure of their small shop in Hospital Street, Perth, in the 1970s. The lower reaches of the Nursery sited to the west of the Dundee Road were provided for use by A K Bell to the people of Perth for allotments in 1920, this being on the eve of national food poverty in 1930. This was converted to greenspace in 1972 and maintained by Perth & Kinross Council. Plans from 1972 indicate that only a small footprint of the original Nursery remained, the assumption being that the majority of the infrastructure was demolished between 1881 and 1920.

Dickson & Turnbull, Hospital Street, Perth
(*W H Findlay Collection, Perth Museum & Art Gallery, Culture Perth & Kinross*)

Street Advert, Perth Rail Bridge, *c.*1895–1912
(*Perth Museum & Art Gallery, Culture Perth & Kinross*)

In 2011, after the closure of the national collection of heathers at Bell's Cherrybank Gardens, Perth, Perth & Kinross Council agreed that a heather collection be developed at Perth's Riverside Park, on the site of the original Nursery's glasshouses. The grounds were again laid out, this time by a landscape architect, employed by the Council. Over 16,000 heather plants, comprising some 500 varieties have been planted by the volunteers of Beautiful Perth who continue to maintain the plantings. The Council maintain the grass and elements of the infrastructure. At the time of writing, Beautiful Perth is planning to establish a Plant Heritage national collection of *Erica carnea* (winter flowering heather).

In 2018 and 2019, the Riverside Park was awarded the prestigious title of RHS Category Winner – Parks and Open Spaces, a discretionary award, awarded to 'that which demonstrated genuine excellence in specific areas'. Fortunately, the true value of the district's greenspaces in terms of physical and mental health and wellbeing is becoming more appreciated, and Beautiful Perth is privileged to have the generosity of the Gannochy Trust and volunteers in allowing it to preserve this precious area for all to enjoy. (This area continues to be leased from the Gannochy Trust on a 99-year lease.)

Riverside Park Heather Collection
(*Elspeth Bruce*)

To the east of the Dundee Road, greenspace which was formerly part of the Nursery remains. It is believed to be owned by the proprietor of Bellwood, which remains as a residential building (with Category B listing). The remaining acreages have been built upon, featuring a variety of architectural styles. However, tree preservation orders exist in the locality protecting a number of trees.

There is much of the Nursery's legacy to celebrate, such as its contribution to commercial horticulture and arboriculture on a national scale, including the development of commercial trading in larch. It is inevitable that many of the wooded landscapes of great estates in Perthshire and beyond would have originated from the Nursery. Perthshire is aptly named 'Big Tree Country', boasting over 200,000 acres of woodland, which includes more champion trees than anywhere else in the UK.

Archibald Turnbull's philanthropic contributions to improvements in Perth, in terms of health, transportation, and the environment were arguably unrivalled by few of his contemporaries. Turnbull's contribution to agricultural and horticultural societies must also be acknowledged, whilst we have Robert Brown to thank for the development of rose breeding, as the foundation of the many beautiful varieties which we are free to enjoy today. It is also inevitable that Perth Nursery would have played a part in David Douglas's choice of career.

Alfred Brown of Alexander & Brown was apprenticed at the Perth Nursery in the 1880s. Alexander & Brown, established in 1897, followed on to become equally notable in terms of their client base and turnover, receiving royal warrants and publishing a beautifully illustrated annual seed and plant catalogue from 1897 until 1963. 'Alf' Brown, a founder of Perth Rotary Club, retired from the business in 1946, on reaching his eightieth birthday and died in 1960. The firm ceased to trade in 1981 when its goodwill and name was bought over by W Smith & Son Ltd, Aberdeen. The firm moved a number of times, but eventually settled in South Methven Street, where the signage 'ALEXANDER & BROWN' is still clearly displayed on the wall above the former shop.

Alexander & Brown, Methven Street, Perth (*Stan Keay*)

The Nurseries are also somehow immortalised in Millais' paintings of *Spring, Autumn Leaves* and *The Vale of Rest*, which now hang in the Lady Lever Art Gallery, Liverpool, Manchester Art Gallery, and Tate Britain, London, respectively. Perth's highly recommended 'Art Trail' also meanders through the former Nursery and includes the Millais frame sculpture with its moulded oak leaves, referencing his *Autumn Leaves* painting and which perfectly frames the spire of St Matthew's Church – the tallest in Perth. William Dickson and Archibald Turnbull are interred in adjacent graves in Kinnoull Churchyard. Scots Roses have been planted in their memory by Beautiful Perth volunteers. Effie Gray is also interred in Kinnoull.

Kinnoull Churchyard, Perth (*Roben Antoniewicz*)

How differently would the great estates and landscapes of today look without the enterprise, sheer hard graft and innovation of these horticultural scions? How much pleasure did their efforts give in appreciating horticulture, silviculture, and arboriculture in all its forms? Their tangible contributions and innovativeness are unlikely to ever be surpassed.

Pomegranate – *Punica granatum*
(*Royal Horticultural Society Lindley Collection*)

Chapter 10

Sketches of the Perth Nursery

T O DATE, no plant catalogues have been discovered by the authors and similarly there appears to be little or no photographic or illustrative evidence of the Perth Nursery, with the exception of one image, believed to be by Perth-based Victorian photographer Magnus Jackson. In the 1870s, there were no fewer than ten professional photographers operating in Perth, yet only one image of Archibald Turnbull has yet been located. Given the level of his public profile, it is highly likely that many more will exist, but are yet to be discovered.

'View of Barnhill, Perth, with Dickson & Turnbull's Nurseries:
Across the River Tay, From the Railway Bridge' by Magnus Jackson, 1880-90
(*Courtesy of Perth Museum & Art Gallery, Perth & Kinross Council, MJS14*)

Botanical illustrations were, and continue to be, extremely popular, but are limited to mostly individual plant species. Advertisements from the 1800s, in for example, *The Gardeners' Chronicle* are valuable in illustrating the gardening equipment, structures, and costumes of the era as well as describing equipment for which the Nursery were agents, such as shading for glasshouses, but none give any illustrations of the Nursery.

Indeed, it is noticeable that the Nursery did not advertise directly as many of its competitors did. It is highly likely that their well-established reputation deemed such expenditure unnecessary. Whilst cartographic records give us an indication of the site of various structures, such as glasshouses, for such a visual subject, words must be relied upon to form a detailed picture of the Perth Nursery:

> Nurseries – So far back as the year 1767, a nursery was formed here, on the east bank of the Tay, by Mr Dickson of Hassendean-burn, in Roxburghshire. He was soon thereafter succeeded by his brother, Mr William Dickson, who, for the long period of 63 years, continued to conduct this extensive establishment in a spirit of enterprise and improvement, and with good taste, which not only made his professional name well known throughout the island, but insured him a wide field of demand for those endless varieties of nature's productions which he is so successful in rearing. Since his death in 1835, its various departments have been conducted in the same spirit, and even on an enlarged scale, under the direction of his nephew, who had for many years taken an active part in its management. These grounds now extend to not less than 60 acres and from their lie, their natural and artificial shelter, and the variety of soils, they contain, it is difficult to conceive a situation more adapted, whether by nature or by art, for every possible purpose to which such establishments are sought to be appropriated. Its giving employment to between 70 and 80 individuals.

> 'Perth Nurseries' in *The New Statistical Account of Scotland* (1845)

The adjoining nursery, now the property of Archibald Turnbull Esq., has been famed throughout Britain during the last half century, for the culture of fruit and forest trees, ornamental shrubbery, and flowers in endless variety. Vast quantities of

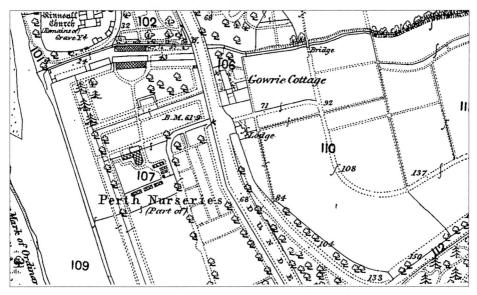

Plan of the Perth Riverside Nurseries, 1884

P.S. *Perth* off Dundee
(*Courtesy of University of Dundee Archives*)

plants are annually sent from this nursery to England. The late
Duke of York, when laying out his pleasure grounds, procured the
greatest number of his ornamental shrubbery from this quarter.

George Penny, *Traditions of Perth* (1836)

Filling up the slope of the hill level grounds for a large extent, on
the right bank of the river, are the beautiful nursery grounds of
Messrs Dickson and Turnbull. These gentlemen, who have
realised a fine fortune from their profession, have at present
upwards of fifty acres in garden cultivation, which are, with a
liberality that characterises all their proceedings, thrown open
as promenade grounds to the public.

The Steamboat Companion betwixt Perth and Dundee (1838)

The nursery grounds of Messrs Dickson and Turnbull, at Perth,
have long been well known to cultivators. They were established
in 1766, and from them have emanated many new and useful
plants. Amongst the flowering plants may be mentioned the first
set of double Scots roses, which were so much in fashion about
forty years ago, but which are now, I regret to find, scarcely ever
asked for. The original stools of these roses still exist, capable of
producing many thousand plants should they be required. These
roses were admirably adapted for the climate of Scotland, and if
they were again in hand by our southern friends, and received a
little of the attention paid to other garden varieties by hybrid-
isation, a new stock might be procured suitable for this country.

James McNab, *Botanical Society of Edinburgh* (1874)

'Services rendered to Forestry by the Perth Nurseries – Their
Formation and Situation – Cultivation of Forest and Ornamental
Trees – The Latest Importations suitable for the Scottish Climate.'

It is undeniable that Perthshire is, to a very large extent,
indebted to the Perth Nurseries for the splendid timber now

existing in its woods and forests; and it is only proper that we should commence our descriptive sketches with an account of those grounds where so many of the trees which now adorn the county have been trained. An old writer, referring to the magnificent wood on Kinnoull Hill, incidentally remarks that to the Perth Nurseries 'the whole vicinity of Perth owes, in a prime degree, the wealth and exuberance of plantation and shrubbery which so extensively beautify it'. This is not only true as regards the neighbourhood of the city of Perth, but it might be applied with equal force to the county generally, and, indeed, to a large part of Scotland, both north and south of Perthshire. Supposing the Perth Nurseries had done no more than brought into existence the arboreal wealth which embellishes the neighbourhood of the 'Fair City', they would have accomplished something to be proud of, and, would have deserved an honourable place in any history of the woods and forests of Perthshire. Without the waving forests that surround the town on every side, – without the magnificent ornamental trees that grow even within its borders, – Perth would not have been entitled to the designation of the 'Fair City'. The everlasting hills, which look down upon it from 'a' the airts the wind can blaw,' would no doubt have enveloped with a rugged grandeur, even although unadorned with growing timber; but how naked and barren would have been the prospect! The great charm of the magnificent view to the north from the centre of Perth Bridge is due to the luxuriant foliage that skirts the eastern bank of the river, and extends as far into the policies of Scone as the eye can reach; while to the south the eye is gratified by the refreshing sight of the dark green pines which clothe the sides and summit of Kinnoull, and by the variegated clumps and isolated trees which everywhere enrich the landscape. What, need we ask, would even Kinnoull Hill be without its plantations of pine? These are some of the arboreal features which surround Perth; and for these we are indebted, in a prime degree, to those Nurseries we propose to describe. It must be borne in mind, however, that what they have done for the immediate surroundings of the city is absolutely insignificant compared with what they have done for the county and the country at large.

Entering the Nurseries by the Bellwood approach, we gradually ascend the side on the hill, and enjoy the invigorating atmosphere, redolent of the ozone from the flourishing pines of Kinnoull. Here, we note, that the Nurseries are not only remarkable for the beauty of their situation, but also for the utility of the site for planting purposes. Lying on the northern and western sides of the Hill of Kinnoull, they are fully exposed to all the cutting blasts, which impart to the young trees that hardiness and vigour so necessary for our Scottish climate. The soil is variable. In some places a nice, friable loam prevails, and in other places the ground is sandy, and occasionally stiff-clay. In many places there are but a few inches of soil, with whinstone cropping up, here and there, above the surface. This variety is most serviceable for practical experiments, and is of great advantage to intending planters. We noticed that in the loam and sandy ground the trees had a most thriving appearance, and those that were lifted carried large masses of fibrous roots.

With an enterprise worthy of the reputation of these grounds, the present proprietor, Mr John A. Anderson, – who has been associated with the Nurseries for nearly half-a-century, and has been enabled to do more to find out that truly hardy varieties of trees of all sorts than probably any other person living, – has endeavoured to secure selections of all the latest importations, so as to be able, from actual cultivation, to speak with confidence of their respective merits. It is satisfactory to learn that most of the importations from the countries named have stood the test of the recent severe winters unscathed.

Thomas Hunter, *Woods, Forests and Estates of Perthshire* (1883)

Appendices

Appendix 1: *Fingask Castle Papers – Plants Bought by Fingask from Dickson & Turnbull, 1855-64*

Appendix 2: *Fingask Castle Papers – Fingask Yearly Amounts Spent with Dickson & Turnbull, 1855-64*

Appendix 3: *Stuart/Stewart of Annat Papers – Tree and Shrub Purchases, 1802-14*

Appendix 4: *Stuart/Stewart of Annat Papers – Accounts/Payments to Dickson & Brown*

Appendix 5: *Stuart/Stewart of Annat Papers – Fruit and Vegetables*

Appendix 6: *The Royal Perth Golfing Society and County & City Golf Club Papers, 1824-90 – From the Minute Books*

Appendix 7: *Memoirs of the Caledonian Horticultural Society Volume 1*

Appendix 8: *The Gard'ners Kalendar*

Appendix 9: *26 George Street, Perth – Shop Premises of Dickson & Turnbull*

Appendix 10: *The Royal Horticultural Society of Perthshire*

Hollyhock – *Alcea rosea*
(*Royal Horticultural Society Lindley Collection*)

Appendix 1:

Fingask Castle Papers
Plants Bought by Fingask from Dickson & Turnbull, 1855-64
Notebook – Dickson & Turnbull Fingask Castle Garden 'Pass Book'
(MS169/3/1/11/171, Perth & Kinross Archive)

8 February 1855
12 Select Hollyhocks
1 g. Laing's Swedish Turnip

15 March 1855
2 g. Mignonette
1 g. New Large Flowering
Mignonette
4 Humea Elegans – 6/- each
[*Incense Plant/Plume Bush now
named Calomeria amaranthoides*]

14 February 1856
12 packets German Asters
10 packets German Stocks
Mixed German Wallflower
2 g. Mignonette
½ lb Sweet Peas
8 packets Flower Seeds at 6d
13 dozen Flower Seeds at 3d
12 dozen Flower Seeds at 2d

19 February 1857
6 packets Globe Quilled German
Asters
6 packets Peony Pyramid Asters
6 packets German 10 Week Stocks
6 packets German Brompton Stocks
3 packets German Wallflowers
2 g. Mignonette

¼ lb Mixed Sweet Peas
1packet Finest Hollyhocks
1packet Carnation
1packet Calceolaria
1packet New [*Text unclear.*]
Spotted Calceolaria

February 1858
Similar list to 1857 with the
addition of 12 Liliums of sorts.
12 Bulbicodium Vernum [*A dwarf
bulb, spring Meadow Saffron,
resembling an autumn flowering
Colchicum, flowers March-April*]
6 Narcissus Bulbicodium [*Hoop
Petticoat Daffodil*]
[*Nasturtium was ordered but was
in with Mustard and Lettuce rather
than the flower orders.*]

1859 Order similar to 1858

Lily – *Lilium tigrinum*
(*Royal Horticultural Society Lindley Collection*)

Lady's Purse – *Calceolaria*
(*Royal Horticultural Society Lindley Collection*)

24 February 1860

1 g Pot Marigold
1 packet Antirrhineam
[*Antirhinnum*] Brilliant
1 packet Dianthus Heddewigii
[*Varieties still available today –
most showy is Black and White
Minstrels.*]
1 packet Primula Sinensis Crem
[*Variety unclear – 2/6 so more
expensive than most.*]

February 1861

8 Papers Chinese Asters
8 Papers German Stocks
Wallflower
Sweet Peas
Mignonette
1 Paper New Primula
6 Papers Poppies

February 1862

1 packet Crystal Palace Nasturtium
Mignonette
Sweet Peas
Asters
German Stocks

April 1862

2 Papers Texican [*Texan*]
Sunflowers

May 1863

1 Adiantum Capillus Veneris
[*Southern Maidenhair Fern*]
November 1863
1 Acacia Hispidissima [*Yellow
Acacia Australia*]
2 Ericea [*Unclear*]
1 Camilla Glauca [*sic*]
1 Camilla Variegata [*Camellia*]
1 Veronica Andersonii Variegata
[*Hebe – still listed.*]
1 Oxylobium Pultenea [*Shaggy Pea
– Australia.*]
1 Helichrysum Barnesii
1 'Statice Halfudii'

February 1864

4 packets Mixed sters
8 ounces Red Prize Quilled Asters
Great Emperor Asters

April 1864

1 packet Ice Plant
Hunt Sweet William
Dream Sweet William
1 packet 'Cocenia'
Sundries [*Long list similar to
Annat. Fingask purchased lots of
tobacco for use as a pesticide.*]
Gishurst's Compund [*An
insecticide.*]
Sulphur, plant fungicide
Rickwett's Compound/Dickwett's
Compound
Shreds and 'Roonds'

NEW SWEET PEAS
1. H.M. STANLEY. 2. M^{RS} ECKFORD. 3. ORANGE PRINCE. 4. DOROTHY TENNANT.

Sweet Peas – *Lathyrus odoratus*
(*Royal Horticultural Society Lindley Collection*)

Appendix 2:

Fingask Castle Papers
Fingask Yearly Amounts Spent with
Dickson & Turnbull, 1855-64

Notebook – Dickson & Turnbull Fingask Castle Garden 'Pass Book'
(MS169/3/1/11/171, Perth & Kinross Archive)

Acreage of Fingask Garden 1868 'Original Garden' 2 Acres & 185 *[roods]*.
Addition 2 roods & 27 falls *[scrap of paper]*.
All of the transactions for plants and sundries in the 'Pass Book' were
signed as paid on or about the second date each year.

	£	s	d
1 December 1854 — 7 December 1855	13	7	6
8 February 1856 — 5 December 1856	11	14	0
19 February 1857 — 11 September 1857	7	2	1
13 November 1857 — 3 December 1858	18	11	6
3 December 1858 — 18 November 1859	13	5	8½
4 November 1859 — 5 October 1860	15	5	0
30 November 1860 — 15 November 1861	14	7	1
14 November 1861 — 30 September 1862	10	7	9½
26 September 1862 — June 1863	11	4	1
20 November 1863 — 2 September 1864	18	13	5
8 November 1864 — 27 October 1865	12	12	4
17 November 1865 — 16 November 1866	10	18	1

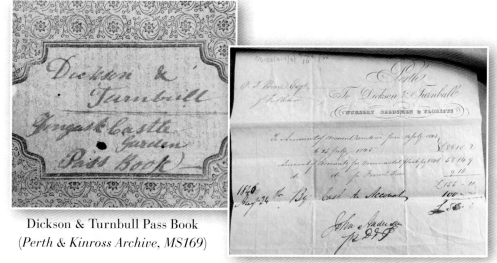

Dickson & Turnbull Pass Book
(*Perth & Kinross Archive, MS169*)

Dickson & Turnbull Receipt

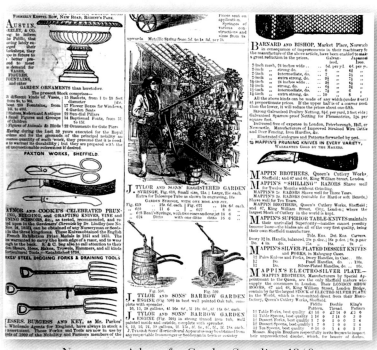

Newspaper Advert for Garden Supplies, 1840

Appendix 3:

Stuart/Stewart of Annat Papers
Tree and Shrub Purchases, 1802-14

*Some examples of the massive number of tree and shrub purchases
by the estate, 1802-14.*

(MS115/Bundles 7 and 8, Perth & Kinross Archive)

Trees and shrubs were being bought on a massive scale –
the list below offer some examples.

Thousands of Trees	*Sundries*
500 Elms	Flowerpots
2,000 Thorns	Brick Wall Nails
6,000 Thorns	Large Wall Nails
100 Spruce	Small Wall Nails
100 Silver Fir	Garron Nails
1,000 Larch	Double Flooring Nails
500 Oaks	Wall Hammer
100 Balm of Gilead Firs (1½ – 2 feet)	Strong Picks
200 Mountain Ash (3 feet)	Quarry Picks
200 Limes (3-4 feet)	Strong Garden Spade
100 Laburnum (3 feet)	Diamond Pointed Shovels
42 Walnuts (1½ feet)	Patent Dutch Hoes
100 Poplars (3-4 feet)	Wood-headed Rake
2,000 Beech (1-1½ feet)	Hedge Shears
50 Balsam Poplar	Peach Pruning Knife
100 Norway Maples (3-4 feet)	Screw Pruning Knife
50 Service (2 feet)	Folding Knife
100 Elders (2 feet)	Sheathed Pruning Knife
200 Roans (*sic*) (3 feet)	Scythe Stones
Swedish Maples	Blue Scythe Stones
Black Italian Poplars	Prime Steel Scythe
Red Cedar	Super Label'd Scythe
White Cedar	Paring Iron
Tulip Tree	Watering Pans

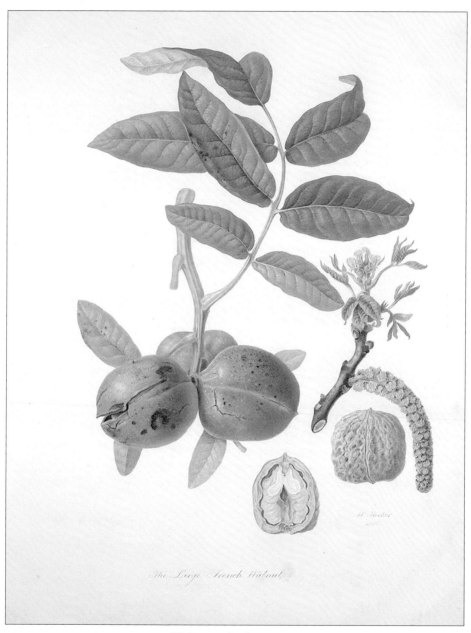

Walnut – *Juglans regia*
(*Royal Horticultural Society Lindley Collection*)

Watering Squirt
Twine
Garden Line
Netting, Pair 'Shama' [*Perhaps
Chamois.*] Gloves
Buckskin Gloves
Bushel Bag
Canvas Bag
Bass [*Coarse Straw*] Matt
Rat Trap
Mole Traps
Stone 'Shreed'
'Roonds' [*Many yards at different
times.*]

Ornamental Plants
72 Varieties of Flower Seeds
38 Varieties Perennial Flower Seeds
Erica Herbaceous
Double Heaths
'Collina' [*Perhaps Calluna.*]
White Jasmine
Ayrshire Roses
Virginia Creeper
Passion Flower
Common Viburnum
Striped Ivy
Clematis Flammula
Evergreen Honeysuckle
Rose Accacea
Rhodendron Ponticum
Azaleas
Budleya Globosa
Laurel Bays
Arbutus; Gum Cistus
Swedish Juniper

Striped Holly
Lilacs
Carolina Bird Cherry
Cotton Lavender
Daphne Pontica
Candleberry Myrtle
Butcher's Broom

Fruit
[*A wide variety and type.*]

Vegetables
[*A wide variety and type of
turnips.*]
Up to 1810:
Yellow Turnip
White Dutch Turnip
Early Dutch Turnip
Yellow Dutch Turnip [*A single
mention on 23 May 1808.*]
1814:
Yellow Aberdeen Turnip
16 Kinds of Tender Annuals

Herbs
Curled Cress
Plain Cress
Curled Parsley
Plain Parsley
Sweet Chervil
Coriander
Fennel
Pot Marjoram
Sweet Marjoram
Summer Savory
Dill
Indian Cress

Garlick
Mustard
Thyme
Purslane
Rosemary
Clary

Seeds
Perennial Ryegrass
White Clover
Red Clover
Yellow Clover

Exotics
Rocambole [*Sand Leek*]
Andromeda
Smilax
Gemma Tamarisk
Colutea
Carolina Allspice
German Tamarisk
Pomegranate

Billing
General Stewart of 'Raitt' to Dickson & Brown, 24 September 1811.
To Seeds for Glendoick & Nursery Plants for 'Raitt' per Account given in commencing 29 September 1807 and ending 23 August 1808: £47–15–11.

To Seeds and Nursery for 'Raitt' per Account given in commencing 13 September 1808 and ending 8 August 1809: £22–12–11.

To Seeds and Nursery for 'Raitt' per Account given in commencing 27 October 1809 and ending 8 August 1810: £6–1–9 / £76–9–9.

Paid Dickson & Brown 23 August 1808: £38–3–9.
Perth, 24 September 1811

'Sir

We take the liberty to send as above a state of your accounts preceding the 1st of September 1810 – the particulars of which were given in to the late William Ross and Mr Thos. Whitson – As we have at present several large items to pay – have the goodness to give instructions to your Man of business to settle them which will very much oblige.

Sir
Your most obedient Hum. Servt. Dickson and Brown.'

———————————————

Appendix 4:

Stuart/Stewart of Annat Papers
Accounts/Payments to Dickson & Brown
Some Examples, 1805-13.
(MS115/Bundles 7 and 8, Perth & Kinross Archive)
25 Feb 1805–2 April 1806
List of thousands of trees of various sizes and species
running to a full page and over the back.

24 September 1811
General Stewart of Raitt to Dickson & Brown.
To Seeds for Glendoick & Nursery Plants for Raitt per account
commencing 29 September 1807 and ending 23 August 1808: £47 16s 11.

To Seeds and Nursery for Raitt per Acct. given in commencing September
1808 and ending 8 August 1809: £22 12s 11d.
To Seeds and Nursery for Raitt per Acct. given in commencing 27 October
1809 and ending 8 August 1810: £6 1s 9d.
Total £76 9s 9d.

23 August 1808
Paid Dickson & Brown: £38 3s 9d.

Edinburgh 15 December 1810
Received from Pat. Murray Threipland Esq of Fingask:
£121 13s 9d payable at Martinmas 1811.
£164 10s 0d payable at Martinmas 1812.
£172 0s 0d payable at Martinmas 1813.

All of which being paid by him at their respective dates is in full of
his Account to us.
Dickson & Turnbull Seedsmen in Perth.

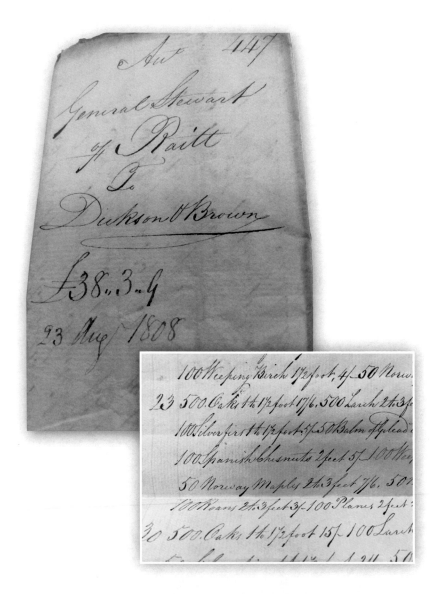

Lieutenant-General Stuart of Annat and Rait Papers
(*Perth & Kinross Archive, MS115*)

Stuart/Stewart of Annat Papers
Fruit and Vegetables
Some Examples
(MS115, Perth & Kinross Archive)

Fruit
Apples on Paradise Stocks
Mulberry
Dwarf Almond
Ryder Apples
Standard Apples
Pears
Cherrys [*sic*]
Dwarf Apples
Dwarf Pears
Dwarf Cherrys
Dwarf Plums
Dwarf Apricots
Ryder Peach
Ryder Nectarines
White Raspberries
Antwerp Raspberries

Vegetables
Apples Dwarf Marrow
Flag Leek
London Leek
Plain Kail [*sic*]
Chardoon
French Endive
Green Curled Endive
Capsicum
Asparagus
Kentucky Beans

Black Speckled Beans
Purple Speckled Beans
Kidney Beans
Long Pod Beans
Scarlet Beans
Early Charlton Peas
Blood Red Beet
White Beet
Marbled Rooted Beet
Early York Cabbage
Late Cabbage
German Green Cabbage
Dutch Cabbage
Green Savoy
Round Leaved Spinach
Cauliflower
White Broccoli
Portsmouth Broccoli
Italian Broccoli
Siberian Broccoli
Silverskin Onions
Strasbourg Onions
Rending Onions
Portugal Onions
Scarlet Onions
Globe Onions
Welsh Onions
Shallots

Garlick
Long Red Carrot
Long Orange Carrot
Early Horn Carrot [*Still available.*]
Yellow Turnip
Early Dutch Turnip
White Dutch Turnips
Yellow Aberdeen Turnip
Swedish Turnip [*Single mention on
23 May 1808.*]
White Cos
Green Cos
Grand Admiral Cos
Curled Cress; Plain Cress
Lamb Lettuce
Curled Parsley
Plain Parsley
Sweet Chervil
Coriander

Fennel
Sweet Marjoram
Summer Savory
Dill
Pot Marjoram
Thyme
Purslane
Clary
Early Radish
Mustard

1814

1 lippie Ash Leaved Potatoes
[*A lippie is a measure of dry goods
– half a gallon.*]
1 lippie Royal Dwarf Potatoes
1 lippie Biggarr Early Potatoes
Magnum Bonum Lettuce
Marrowfat Pease
Scotch Leek

Appendix 6:

The Royal Perth Golfing Society and County & City Golf Club Papers, 1824-90
From the Minute Books
(GB/252/MS184, Perth & Kinross Archive)

21 April 1824
Opening Meeting: Archibald Turnbull, Seedsman is recorded as a founding member. Election of office bearers – Archibald Turnbull Esq., 2nd Councillor.
Committee consisted of four councillors, a secretary and treasurer.
The club moved to the North Inch in 1834 – 10 holes.
Club Uniform: 'Scarlet frieze with dark green collars and white metal plain plated buttons'.

21 April 1840
Mr Archibald Turnbull moved an amendment that there should be a new 'Button' bearing an insignia – this proposal was later rejected in favour of the plain one.

14 October 1874
Autumn AGM: Archibald Turnbull was present – only three months before his death.

Appendix 7:

Memoirs of the Caledonian Horticultural Society – Volume I
Prizes Proposed for the Year 1812

I.   **The Production of Fruits and Culinary Vegetables, and Flowers. (The Prize the Society's Silver Medal, or a Guinea and a Half, in the Option of the Gainer).**

To be shown at the Quarterly Meeting on the second Tuesday of March, or to the Committee at the Physician's Hall, on the second Tuesday of May.

1.   The best brace of early Cucumbers;seeds to be sown in January.
2.   The best cluster of Early Grapes, of any kind.
3.   The best six heads of Spring Broccoli.
4.   The best six stems of Brussels Sprouts.
5.   The best six heads of Winter Lettuces.
6.   The best six seedlings Polyanthuses; from seed sown last year.
7.   The best early Melon.

To be shown at the Quarterly Meeting on the second Tuesday of June, or to the Committee at the Physicians Hall, of the dates after mentioned.

1.   The best Melon.
2.   The best six forced Peaches, with names.
3.   The best six heads of Cauliflower
4.   The best three seedling Pinks, from seeds sown Summer 1811 – 2nd Tuesday of July
5.   The best twelve sorts of Gooseberries, twelve berries of each sort, with their names – 1st Tuesday of August.

To be produced at the Quarterly Meeting on the second Tuesday of September.

1.   The best six Peaches, from the open air.
2.   The best six Nectarines, from the open air.
3.   The best six Apricots with names.
4.   The best dozen of Green Gage Plums.

5. The best six Jargonelle Pears.
6. The largest cluster, and best swelled berries, any kind of Grapes.
7. The best six seedling Carnations.
8. The best Home-made Wine, without the use of any imported material excepting sugar; two bottles to be produced, together with an account of the method of preparing it.

To be produced at the Quarterly Meeting on the second Tuesday of December

1. The best six kinds of Apples, three of each sort, with their various local names, particularly kinds not generally known, with their peculiarities and history.
2. The best six kinds of Pears, three of each sort, also with names &c.
3. The best six heads of Late Broccoli.
4. The best six heads of forced Sea Cale
5. The best twenty-five heads of forced Asparagus

II. The Production of New or Improved Varieties of Fruits, Culinary Vegetables or Flowers.

1. For the best new Apple, adapted to the climate of Scotland, raised from seed. Ten years to be allowed. Gold medal and Twenty Guineas.
2. For the best new Pear, raised as above. Ten years to be allowed Gold Medal and Twenty Guineas.
3. For the best new Peach or Nectarine, raised as above. Six years to be allowed. Gold medal and Ten Guineas.
4. For an improved variety of Dutch Currant, raised from seed. Five years to be allowed. Gold Medal, or, in the option of the gainer, Five Guineas and the same commutation will be allowed in other cases.
5. For the best new Early Cucumber. Gold Medal.
6. For the best new Strawberry, raised from seed. Four years to be allowed. Gold Medal.
7. For the best new sort of Early Potato, *without blossoms*, raised from seed. Five years to be allowed.

Pêcher à fruits lisses.

P. J. Redouté   Victor

Nectarine – *Prunus persica var. 'nucipersica'*
(*Royal Horticultural Societ Lindley Collection*)

III. Communications &c (The Gold or Silver Medal to be Awarded by the Committee, According to the Value and Importance).

1. On the best method of improving the sorts of Broccoli already cultivated, and of saving their Seeds genuine in this climate.
2. The best method of cultivating and forcing Sea Cale.
3. The best treatise on Orchard Fruits adapted to the climate of Scotland, with lists and descriptions of the different kinds – their habits of growth, &c – their synonyms or local names; those for the table, and those for kitchen use.
4. The best treatise on the culture of the Dutch Currant for wine.
5. The best mode of preventing or curing the Mildew upon different Fruit-trees and other Vegetables.
6. The best mode of preventing or curing the Canker in fruit trees, &c.
7. The cheapest and most effectual mode of preserving Fruit trees on walls form the effects of spring frosts.
8. The best mode of destroying the blue insect, breeding in the crevices of the bark of Apple trees, causing them to canker and die, chiefly on those trees imported from the London nurseries.
9. The best method of destroying Wasps, Wood-lice, Earwigs &c infesting wall-fruits.
10. The best method of preventing Worms in Carrot, in Cauliflower, and Broccoli roots.
11. The best mode of destroying the Wire-Worm.
12. The best mode of destroying the Pine-bug, the brown scale, the white bug, the Aphis or Green fly – the Chermes – the Red Spider – the Thrips; or any other insect infecting Hot-houses, Pits, Melon and Cucumber frames &c.
13. The best means of increasing the quantity of Manure, and the best mode of applying it to different crops.

Appendix 8:

The Gard'ners Kalendar*

*Shewing the most seasonable times for performing his Hortulan Affairs
Monthly throughout the Year: And a Catalogue of such dishes and drinks
as a compleat Garden can afford in their season,
by JOHN REID, Gard'ner.*

Reader,
As in this little kalender thou wilt find when; so in my book (intituled
The Scots Gard'ner) thou wilt find how, to perform the particulars. The
gard'ners year is a circle as their labour, never at an end.

Nevertheless their terme is.

January

Prepare the grounds, soils and manures. Fell trees for mechanical uses.
Prune firrs, plant hawthorn hedges and all trees and shrubs that lose the
lead if open weather. Also prune the more hardie and old-planted. Manure
the roots of the trees that need. Drain excessive moisture; gather graffs ere
they sprout, and near the end graff. Begin wih the stone fruits. Gather
holly berries, firr husks, &c. Secure choice plants as yet from cold and wet,
and earth up such as the frosts uncovered.

Feed weak bees, also you may remove them.

Garden Dishes and Drinks in Season
Colworts, leeks, &c. Dry sweet herves, housed cabbage, onions, shallot,
parsneeps, skirrets, potatoes, carrots, turneeps, beet-rave, scorzonera,
parslet and fennel roots in broth.

Pickled artichocks, beet-raves, &c. Housed aples, pears, and other
conserved fruits.

Cyder and other wines as before.

---

* The excerpt from this book appears for historical record only and the practices within it
are not advised for the modern gardener, especially in regard to the destruction of wildlife.

February

Plant any trees or shrubs that lose the leaf, also lay such for increase; see *June.* Likewayes sow all your seeds, kyes, kirnells, nets, stones; also the seeds of several greens, as holly, yew, philyrea, laurels, &c. Prune firrs,&c. Continue to destroy vermine.

Graffing is now in season, see the last moneth.

Prune all trees and shrubs except tender greens.

Nail and dress them at the wall. Cover the roots of trees layed bair the fore-end of winter, if any be. Plant hawthorn hedges, willows &c.

Plant liquorish, potatoes, peas, beans, cabbage, sow parsley, beets, spinage, marigold and other hardie pot-herbes.

Let carnations and such sheltered flowers get air in mild weather. But keep close the green-house.

Now you may remove bees and feed weak stocks.

Garden Dishes and Drinks in Season

Cole, leiks, sweet herbes, onions, shallot, housed cabbage, skirrets, turneeps, pars-neeps, potatoes, beet-rave, scorzonera, carrots, besides parsley and fennel roots.

Pickled beet-rave, artichock, cucumber; housed aples, pears, and other conserved fruits with cyder and other wines and drinks, as above.

March

Re-delve, mix and rake your ground for immediate use. Delve about the roots of all your trees. Yet plant trees and rather greens. Also prune such except resinous. Propagate by laying, circumposition, and especially by cuttings. Sow the seeds of most trees and hardie greens. Cover those tress whose roots lay bair and delve down the manures that lay about your young trees all winter, covering on leitter again topt with earth to prevent drought in summer: this is a material observation and more especially for such as are late planted. Slit the bark of ill-thriving trees. Fell such as grow croked in the nurserie. Graffing is yet in season (but too late for stone fruit), cut off the heads of them inoculated.

Set peas, beans, cabbage. Asparagus, liquorish. Sow parsley, beets, endive, succory, bugloss, burrage, seller, fennel, marigold. Plant shallot, garleeks, potatoes, skirrets. Sow onions, lettice, cresses, parsneeps, beet-rave, radish &c. and on the hotbed coleflower, and if you please cucumber, &c.

Slip and set physick herbes, July-flowers, and other fibrous-rooted flowers. Be careful of the tender plants' the piercing colds are now on foot. Turn your fruit in the room but open not yet the windows.

Catch moles, mice, snails, worms, destroy frogs spawn, &c.

Half open passages for bees, they begin to flit; keep them close night and morning: yet you may remove them.

Garden Dishes and Drinks in Season
Both green and housed herbes and roots: also pickled, housed and conserved fruits, with their wines as in the former months.

April
Plant holly hedges and hawthorn too, if not too foreward. Ply and sheer hedges. Nail an prune wall-trees, &c. Sow and plant firrs, and other greens. Slip and set sage, rosemary, thym, rue, savoury and all fibrous rooted herbes and flowers. Uncover and dress strawberries.Plant artichocks, slip them and delve their plottes. Set cabbages, beans, peas, kidnees. Sow asparagus, parslet, beets and beet-card. Set garleeks, shallot, potatoes, skirrets, sorral. Sow onions, leeks, lettice, cresses, radish, orach, scorzonera, carvy, fennel, &c. And the hot-bed, cucumbers, coleflowers, purslain, sweet marjoram, basil, summer savoury, tobaco,&c.

Set strawberries, violet, July-flowers, &c. Also sow the seeds of July-flowers, &c. Sow all your annuall flowers and rare plants, some requiring the hot-bed. Lay, beat, and roll gravel and grass. Fall to your mowing and weeding.

Destroy moles, mice, worms and snails.

Open the doors off your bee-hives, now they hatch.

Garden Dishes and Drinks in Season
Onions, leeks, colworts, beets, parsley, and other herbes: spinage, sorral, scorzonera; green asparagus, lettuice, and other sallads. Pickled artichocks, beet-rave, barberries, cucumbers.

Housed aples and pears, conserved cherries,plumes, peaches, apricocks, goosberries, currans. Also wines of aples, pears, cherries, liquorish, hony, &c.

ALBUM BENARY.
Tab. XXIII
gr. nat.

Ad. nat. pict. in horto. Benary.

Chromolith. G.Severeyns, Bruxelles.

ERNST BENARY, ERFURT.

Sugar Peas – *Pipsum sativum*
(*Royal Horticultural Society Lindley Collection*)

May

Pull up suckers and haw about the trees. Rub off unnecessary buds. Sheer or clip hedges. Prune tender greens (not the resinous), bring furth the houses ones refreshing and trimming them. Plant all sorts of medicinal herbes.

Sow all sweet ones which are tender.

Garther snails, worms and catch moles.

Sow lettice, cresses, purslain, turneep, radish, peas, &c. Continue weeding and watering.

Near the end watch the bees ready to swarm

Garden Dishes and Drinks in Season

Coleworts and other herbes, (being eaten with contentement are better than a fatted ox without it), sage (with butter), leeks, parsley, thyme, marjorum, sorrall, spinage, &c. Scorzonera, asparagus, lettice, purslain, and other sallades and pot herbes.

Pickled artichocks, barberries, beet-rave, cucumbers, housed aples and pears for many uses. Early cherries, strawberries, near the end. Cyder, metheglin, liquorish ail, &c.

June

Cleanse about the roots of trees, suckers, and weeds; water their covered bulks, especially the new planted.

Fell the long small ill-train'd forest-trees in the nurserie, within half a foot of the ground. Unbind graffs. Prune all wall and standard trees. Towards the end you may inoculate and also increase by circumposition.

Gather elm seed and sow immediately. Transplant coleflowers, coleworts, beets, leeks, purslain, &c., in moist weather; at te least water first the ground if dry.

Sow peas, radish, turneep, lettice, chervil, cresses, &c.

Destroy snails, worms, &c.

Begin to la carnations or July-flowers; shade, support and prune such as will blow. Water the pots and thirsty-plants. Weeding and mowing is in season, and so is distillation.

Bees now swarm, look diligently to them.

Garden Dishes and Drinks in Season
Cole, beets, parsley, sorrall and other pot-herbes. Purslain, lettice, and
other sallads. Radish, scorzonera, asparagus,green pears and artichocks.
Green goosberries. Ripe cherries, rasps, currans, strawberries.

Housed aples and pears.

Cyder, metheglin, &c.

July

Fallow ground as soon as the crop comes off. Prune and purge all standard
trees. Ply, nail, prune, and dress your wall-trees. Pull up suckers and
weeds. Haw and water where needful. Inoculate fruit-trees, shrubs, rare
greens, and flower-trees; increase the same by laying. Clip your hedges
after rain. Suffer such herbes and flowers to run to seed as you would save,
cutting the rest a handful from the ground.

Sow turneep, radish, lettice, onion, cole-flower, cabbage, and coleworts
in the full moon. Near the end sow beets, spinage &c. You may plant
strawberries, violets, camomile. Lay July-flowers. Plant their seedlings. Slip
and set hypaticas, bears-ears, couslips, helibors, &c. take up bulbous and
tuberous ones that are dry in their stalks (if you mind to change their
places) and keep till September, but some should be set immediately.

Supply voids with potted annuals. Lay grass and gravell. Make cherrie
and raspberrie wine, &c.

Prevent the bees' later swarms Kill drons, wasps, &c.

Garden Dishes and Drinks in Season
Beets and many potted herbes and sweet herbes.

Beet-card, purslain, lettice, endive, &c.

Cabbage, cole-flower, scorzonera, beet-rave, carrot, radish, turneep,
peas, beans, and kidness, artichocks, strawberries, rasps, currans,
goosberries, cherries, plumes, summer pears and aples.

Cyder, metheglin, and other wines.

August

Fallow bordures, bed, nurseries, and the bulks of trees. Yet inoculate. Ply
and purge trees. Pull up suckers and weeds. Clip hedges. Gather the stones
of black cerrie and morella. Gather mezerion berries. Gather the seeds of

most herbes and flowers. Cut your physick herbes. In the beginning sow cabbage (tho'I confess it's too late. See last moneth). Beets and beet-card, spinage, black radish, chervil, lettice, corn-sallade, endive, scorzonera, carvy, marygold, angelica, scurvy-grass, &c. Take up ripe onions, garleeks, and shallot. Unbind buds inoculated.

Cut and strong strawberries. Lay July-flowers. Sow columbines, holyhoks, lark-heels, candy tuffs, popies, and such as can endure winter.

Take up you bulbs and plants as in the last. Sift the ground for tulips and glasiolus. Plunge in potted annuals in vacants. Keep down weeds by hawing. Lay grass, beat, roll and mow well. Make gooseberries and curran wine.

Towards the end take bees, take the lightest first; those that are near heaths may differ a little. Destroy wasps, straiten the passage by putting on the hecks to secure from robers.

Garden Dishes and Drinks in Season
Many pot-herbes and sallades, cabbage, cole-flower, beet-card, turneep, radish, carrot, beet-rave, scorzonera, peas, beans, and kidness, artichocks, cucumbers, aples, pears, plumes, apricocks, geens, goosberries, currans, rasps, strawberries, &c.

Cyder, metheglin, cherrie wine, curran wine, goosberrie wine, raspberrie wine, &c.

September
Fallow, trench, and level ground. Prepare pits and bordures for trees. Gather plane seed, also almond, peach, and white plum stones. Gather ripe fruits. Plant furth cabbage. Remove bulbs and plant them. Refresh, trame, and house your tender greens. Refresh and trim pots and cases within July-flowers and other fine flowers and plants; carrying the to pits, shelter and covert, giving them air.

Toward the end gather saffron.Make cyder, perry and other wines.

Straiten the entrance to bee-hives, destroy wasp, &c.

Also you may remove bees.

Garden Dishes and Drinks in Season
Varieties of pot-herbes and sallades, cabbage, cole-flower, peas, beans, and kidness, artichocks, beet-card, beet-rave, scorzonera, carrots, turneeps,

radish, cucumbers, apes, pears, apricocks, peaches, nectarines, quince, grapes, barberries, filbeards.

Cyder, liquorish ail, metheglin, and wine of cherries, rasps, gooseberries, currans, &c.

October
Gather winterfruits. Trench and fallow grounds (mixing proper soil) to ly over winter. Prepare manures, mixing and laying in heaps bottom'd and covered with earth. Plany hawthorn hedges, and all trees that lose their leaves. Also lay their branches. Prunes roses. Gather seeds of hassell, hawthorn, plan, ash, beach, oak, aple, pear, &c.

Cut strawberries, artichocks, asparagus, covering their beds with manure and ashes. Earth up winter sallades, herbes and flowers a little. Plant cabbage, tulips, anemonies and other bulbs. Sow the seed of bairs-ears, cowslips, tulips, &c. Beat and roll gravel and grass. Finish your last weeding and mowing. Lay bair leopered tree roots and remove what harms them; also delve and manure such as require it. Drain excessive moisture wherever it be. Pickle and conserve fruits. Make perry and cyder.

You may now safely remove bees.

Garden Dishes and Drinks in Season
Coleworts, leeks, cabbage, coleflowers, onions, shallot, beans. Blanche endive and sellery, pickled asparagus, purslain,&c.
Scorzonera, beet-rave, carrots, turneeps, parsneeps, potatoes, skirrets, artichocks, cucumbers, aples, pears, plumes, almond &c.
Cyder, perry, and wine of cherries, currans, goosberries, rasberries, ail of liquorish, metheglin, &c.

November
Contrive or forecast where, and what you are to sow and plant. Trench and fallow all your vacant grounds. Prepare and mix soils and composts thoroughly; miss not high-way earth, cleanings of streets; make compositions of manures, soils and lyme.

Lay bair roots of trees that need, and manure such as require it. Plant all fruit-trees, forrest-trees, and shrubs that lose the leaf, also prune such. Plant cabbage, sow hasties for early peas in warme grounds, but trust not to them.

Gather the seeds of holly, yew, ash, &c., ordering them as in Chap III. Furnish your nurseries with stocks.

Shelter tender evergreen seedlings. House your cabbage, carrots, turneeps: and at any time ere hard frosts house your skirrets, potatoes, parsneeps, &c. Cover asparagus, artichocks, as in the last moneth. Sow bairs-ears, plant tulips, &c. Shut the conservatory. Preserve your choisest flowers. Sweep and cleanse the walks of leaves, &c.

Stop your bees close so that you leave breathing vents.

Garden Dishes and Drinks in Season
Cabbage, coleflower, onions, leeks, shallots, &c. Blanched seller, succory, pickled asparagus, purslain, &c. French parsneeps, skirrets, potatoes, carrots, turneeps, beet-rave, scorzonera, parsley and fennel roots, apples, pears &c.

Cyder, perry, wine of cherries, rasps, currans, gooseberries, liquorish, hony &c.

December

Trench and prepare grounds. Gather together composts; plant trees in nurseries, and sow their seeds that endure it.

Gather firr seeds, holly berries, &c. Take up liquorish. Continue your care in preserving choice carnations, anemonies, and ranunculuses from raines and frosts. And keep the green-house lose against piercing colds.

Turne and refresh your fruit in a clear serene day. Sharpen and mend tools. Gather oziers and hassell rods and make baskets in stormy weather. Cover your water pipes with leitter lest the frosts do crak them; feed weak bees.

Garden Dishes and Drinks in Season
Colworts, leeks, &c., housed cabbage, onions, shallot. Several dryed sweet herbes. Houses parsneeps, turneeps, skirrets, carrots, potatoes, beet-rave, scorzonera; parsley and fennel roots. Pickled cucumbers, barberries, artichocks, asparagus, purslain, &c.

Housed aples, pears. Conserved cherries, plumes, peaches, apricocks, &c.
Wines of aples, pears, cherries, liquorish, honey, &c.

Appendix 9:

26 George Street, Perth
Shop Premises of Dickson & Turnbull

Cauliflower – two very superior heads of this excellent vegetable are to be seen in one of the windows of Messrs Dickson and Turnbull, George Street; the larger of the two measures fully 2 feet 8 inches in circumference; the other two feet six inches. They are from the garden of Methven Castle and are assuredly very creditable to Mr Bissett the gardener, whose success in the growing of cauliflower may be conceived, when we say that the couple alluded to form a sample of his general crop.

Perthshire Constitutional and Journal, 12 October 1836

Among the various remarkable specimens of agricultural and horticultural skill so frequently exhibited by Messrs Dickson and Turnbull, our attention was arrested the other day by a most gigantic onion, apparently large enough to season the luncheon of a whole corps of Welsh troopers. The net weight of this immense onion which was grown in the garden of Mr Marshall Luncarty, is 2 pounds 10 ounces and measures 19 inches in circumference.

Perth Courier, 22 November 1836

There is to be seen in the window of Messrs Dickson and Turnbull nurserymen here a shoot of a white Antwerp rasp of this year's growth, 9 to 10 feet long and bearing 72 berries grown in Mr Clark's garden at Craigie. In this climate it is a rare thing to find wood of the years growth producing and ripening fruit and is a proof of the fineness of the past season.

Perthshire Courier, 22 September 1842

There has been for some days exhibiting in the shop window of Messrs Dickson and Turnbull, seedsmen, here, among other fine specimens of agricultural produce, a white globe turnip weighing 25½ pounds imperial. This was raised by Mr William Moncur, Blairgowrie, and we believe is unprecedented in this, and we may suppose in any other quarter. The field was manured with bones and dung in nearly equal proportions, and the whole crop is generally very large.

Sussex Advertiser, 31 December 1844

High Farming – A sample of white oats is now exhibiting in Messrs Dickson and Turnbull's in George Street, grown up on a 20 acre field of alluvial land, lately reclaimed by Sir John S. Richardson, Bart., upon his estate at Pitfour. The oats which is a fair sample of the whole field stand from 5 to 6 feet in the straw; from each root, 25 to 30 stems have sprung spanning from 200 to 230 grains – each seed therefore has multiplied from 5,000 to 7,000 fold. The field was reaped last week and so thick was the crop that there is hardly space for the stooks to stand on the ground. The exact extent of the field is 19½ imperial acres and produce 2,160 stooks of 12 sheaves each.

Dundee Courier, 24 September 1851

Large potatoes – on Friday last we observed, with much interest, a few specimens of very large potatoes in the shop window of Messrs Dickson and Turnbull George Street, which had been taken from a considerable parcel grown on the farm of Murhall. The samples were six in number and weighed in the aggregate 9¼lb, while one of them weighed 33oz.

Morning Advertiser, 24 October, 1855

The Colorado Beetle – Preserved specimens of this destructive insect are exhibited in the window of Messrs Dickson & Turnbull, George Street.

Dundee Courier, 29 July 1877

Appendix 10:

The Royal Horticultural Society of Perthshire

On Friday last the first meeting this Society for the year for competition in Flowers, Fruit, and Vegetables was held in Mrs Wallace's Hall, Athole Street; and, as usual at the meetings of this association, the attendance was numerous and fashionable. From the lateness of the season, the various productions of the garden were rather deficient in quantity compared with some former meeting, but nevertheless of excellent quality – a circumstance which shewed how much may be effected by skill and perseverance, even in seasons so very uncongenial and backward as the past spring.

Amongst many things sent for the purpose of decorating the Hall, we noticed, on the side tables, many beautiful plants, in pots, from the nurseries of Messrs Dickson and Turnbull, and about 83 varieties of apples, and 3 varieties of pears, in good condition, from the same gentlemen. Also, a collection of Hyacinths and Polyanthus Narcissus from Mr Stewart's Nursery. And, from Kinfaun's Garden, some fine plants, of different varieties, of the Cactus, amongst which was a beautiful Hybrid Seedling, called Cactus *Speciosus Robertsonii*, and some ripe Strawberries from the same place. A plate of equally fine Strawberries, from Invermay; and from the garden at Moncrieff, a plant of *Amarylis Jenkensonii*. A beautiful plant of *Erica Aristata* from Delvine Garden...

Perthshire Constitutional and Journal, 8 May 1835

Select Bibliography

PETER D A BOYD, Various publications and articles on Scots Roses, http://www.peterboyd.com/index.htm [accessed August 2020]

GEORGE BUIST, *The Steamboat Companion betwixt Perth and Dundee* (Edinburgh: 1838)

E H M COX, *A History of Gardening in Scotland* (London: 1935)

REVEREND LEWIS DUNBAR, *The Statistical Account of Scotland: Kinnoul County of Perth*, OSA 1796, Volume 18

JOHN FORBES, *Hortenis woburnensis: a descriptive catalogie of upwards of six thousand ornamental plants cultivated at Woburn Abbey.* (London: 1833)

Gardeners' Chronicle and New Horticulturalist (1857)

GRAHAM HARDY, 'Robert Brown of Perth and James McNab's North American tour of 1834', Sibbaldia: *The Journal of Botanic Garden Horticulture*, Number 9 (2011)

HAWICK ARCHAEOLOGICAL SOCIETY, *Transactions* (2010)

THOMAS HUNTER, *Woods, Forests and Estates of Perthshire* (Perth: 1883)

JOHN CLAUDIUS LOUDON, 'Biographical Notice of the late Mr David Douglas', *Gardener's Magazine* (1836)

JOHN CLAUDIUS LOUDON, *The Suburban Gardener, and Villa Companion* (London: 1838)

JOHN CLAUDIUS LOUDON, *Arboretum et Fruticetum Britannicum* (1838)

PATRICK NEILL FLS, *On Scottish Gardens and Orchards, Drawn up, by desire of the Board of Agriculture* (1819)

PATRICK NEILL FLS, *Journal of a Horticultural Tour Through Some Parts of Flanders, Holland and the North of France in the Autumn of 1817 – Transactions of the Caledonian Horticultural Society* (1823)

GEORGE PENNY, *Traditions of Perth* (Perth: 1836)

JOSEPH SABINE, 'Description and Account of the Varieties of Double Scotch Roses cultivated in the gardens of England', *Transactions of the Horticultural Society of London Journal*, Volume 4 (1822)

SUE SHEPHARD, *Seeds of Fortune: A Great Gardening Dynasty* (London: 2012)

T C SMOUT, *A History of the Scottish People, 1560-1830* (London: 1969)

GEORGE TANCRED, *The Annals of a Border Club (the Jedforest)* (Jedburgh: 1899)

The New Statistical Account of Scotland, Volume 10: Perth (Edinburgh: 1845)

TIPPERMUIR BOOKS

Tippermuir Books Ltd (est. 2009) is an independent publishing company based in Perth, Scotland.

OTHER TITLES FROM TIPPERMUIR BOOKS

Spanish Thermopylae (2009)

Battleground Perthshire (2009)

Perth: Street by Street (2012)

Born in Perthshire (2012)

In Spain with Orwell (2013)

Trust (2014)

Perth: As Others Saw Us (2014)

Love All (2015)

A Chocolate Soldier (2016)

The Early Photographers of Perthshire (2016)

**Taking Detective Novels Seriously:
The Collected Crime Reviews of Dorothy L Sayers** (2017)

Walking with Ghosts (2017)

No Fair City: Dark Tales from Perth's Past (2017)

**The Tale o the Wee Mowdie that wantit tae ken
wha keeched on his heid** (2017)

**Hunters: Wee Stories from the Crescent:
A Reminiscence of Perth's Hunter Crescent** (2017)

Flipstones (2018)

**Perth: Scott's Fair City: The Fair Maid of Perth & Sir Walter Scott –
A Celebration & Guided Tour** (2018)

God, Hitler, and Lord Peter Wimsey: Selected Essays,
Speeches and Articles by Dorothy L Sayers (2019)

Perth & Kinross: A Pocket Miscellany:
A Companion for Visitors and Residents (2019)

The Piper of Tobruk: Pipe Major Robert Roy, MBE, DCM (2019)

The 'Gig Docter o Athole':
Dr William Irvine & The Irvine Memorial Hospital (2019)

Afore the Highlands: The Jacobites in Perth, 1715-16 (2019)

'Where Sky and Summit Meet': Flight Over Perthshire – A History:
Tales of Pilots, Airfields, Aeronautical Feats, & War (2019)

Authentic Democracy: An Ethical Justification of Anarchism (2020)

'If Rivers Could Sing': A Scottish River Wildlife Journey.
A Year in the Life of the River Devon as it flows through the
Counties of Perthshire, Kinross-shire & Clackmannanshire (2020)

A Squatter o Bairnrhymes (Stuart Paterson, 2020)

In a Sma Room Songbook: From the Poems by William Soutar (2020)

The Nicht Afore Christmas:
the much-loved yuletide tale in Scots (2020)

The Shanter Legacy (Garry Stewart, 2021)

———

BY LULLABY PRESS
(an imprint of Tippermuir Books)

A Little Book of Carol's (2018)

Diverted Traffic (2020)

———

FORTHCOMING

Ice Cold Blood (David W Millar, 2021)

**Fatal Duty: The Scottish Police Force to 1955:
Cop Killers, Killer Cops & More** (Gary Knight, 2021)

William Soutar: Collected Poetry, Volume I *(Published Work)*
(Kirsteen McCue and Paul S Philippou (eds), 2021)

William Soutar: Collected Poetry, Volume II *(Unpublished Work)*
(Kirsteen McCue and Paul S Philippou (eds), 2022)

A Scottish Wildlife Voyage (Keith Broomfield, 2021)

**Beyond the Swelkie: A Collection of New Poems & Writings to
Mark the Centenary of George Mackay Brown (1921-1996)**
(Jim Mackintosh & Paul S Philippou (editors), 2021)

———

All Tippermuir Books titles are available from bookshops and
online booksellers. They can also be purchased directly
with free postage & packing (UK only) –
(minimum charges for overseas delivery) from
www.tippermuirbooks.co.uk

Tippermuir Books Ltd can be contacted at
mail@tippermuirbooks.co.uk

TIPPERMUIR
· BOOKS LIMITED ·